T0076462

THE SIXTH ELEMENT

CONTENTS

THE SIXTH ELEMENT

Introduction

In a Universe of wonders, carbon is truly a wondrous element. Carbon can be hard or soft, sooty black or clearer than crystal. It was forged in the fiery interiors of stars; in its clear diamond form, it feels cold in your hand; and out of your hand, it is the very best conductor of heat. Burning forms of carbon produced heat that kept humans warm for millennia and energy that powered the Industrial Revolution. We eat tons of it in our lifetime, and the unique chemical prowess of this Swiss army knife of chemical elements forms the virtual backbone of life as we know it. We are made of it, as Joni Mitchell wrote in her song "Woodstock," "We are stardust . . . billion year old carbon."

This book is dedicated to just one of the naturally occurring chemical elements. Carbon is considered the sixth element because it has six protons in its nucleus and six electrons to balance the protons' positive charge. Considering its abundance in the Sun, carbon ranks as the fourth most abundant element after hydrogen, helium, and oxygen. Oddly enough, despite its high abundance in stars, carbon is relatively rare inside our planet. We see lots of it near the Earth's surface, where we live, but averaged over the whole planet, it is actually a rare element. We live in a carbon-rich environment on the surface of a carbon-poor planet. We will explain how this happened and why we are different from many outer solar system bodies in chapter 3.

In this book we also discuss how carbon was discovered and how understanding this important element was a major advancement of our scientific understanding of nature. We will see how it is made in stars, how carbon atoms ended up on Earth, and why it can form so many compounds that are key to our existence. We will explore some of the impacts that the sixth element has had on human history, many of its remarkable uses, and its role in the past and future of our planet.

Like other elements heavier than hydrogen, carbon is just a tiny core of comparatively massive protons and neutrons that is surrounded by electrons. Carbon, however, is unlike all the other elements in its ability to bond with other atoms to make materials with an extraordinary range of chemical and physical properties. When other atoms are involved, a nearly unlimited number of compounds can be made, including some very complex ones that enabled the formation of life and then evolved over geologic time to produce the living organisms that we know of.

Carbon may be just one of nearly a hundred naturally occurring elements, but it stands out from all the others. As a pure element, it can exist in such diverse solid forms as soot, graphite, diamond, buckyballs, nanotubes, and sheets of carbon lattice only a single atom thick. When bonded to other elements, it can form a nearly infinite number of compounds. These compounds are so important that they are granted their own class of "organic chemistry."

There are several known natural forms of pure elemental carbon, and at least one unnatural one. The simplest pure carbon form is just a single atom. This is not found on Earth, because carbon atoms stick to everything and form molecules. Only in the isolation of interstellar space are single-carbon atoms found. Next are carbon chains, carbon atoms in line, which also exist naturally only in interstellar space.

When carbon atoms bond to each other, they can form sheets named graphene. When graphene sheets stack together, which is easy, we have graphite, the stuff that allows pencils to make black lines. Carbon sheets can curl up to form tiny hollow nanotubes. The next in complexity are the fullerenes, named after Buckminster Fuller. The most abundant fullerene in nature is the semispherical molecule C_{60} that resembles a soccer ball. Carbon famously also forms crystals of the superlative mineral diamond, to which we dedicate one chapter. Rings of pure carbon are also possible: cyclocarbon, with eighteen carbon atoms, is the only one that has been made. It was predicted in theory but not produced until 2019. So, even with the incredible diversity of carbon that we do know about, there is always more to discover.

As astronomers, we have inevitably put a broad cosmic focus on the many aspects of carbon. As scientists, we have placed strong emphasis on the fundamental science issues involved with this special element. And because of carbon's incredible role in both the history of humans and science, we have also chosen to view carbon through a lens of history. The broad range of carbon's history involves its origin, how it served as a fundamental gateway to the formation of nearly all of the other chemical elements, how it evolved in space, how it got to Earth, and how it was used to make life and drive the evolution of our planet. Early human dealings with carbon led to making fire and cave paintings, and then evolved into the foundations of science and our first understandings of atoms and what matter actually is. History does not end now, and the findings of both physics and astronomy clearly show the basic roles that carbon will play even trillions of years into the future, as it cycles between vastly different environments and is ultimately destroyed in the difficult-to-fathom deep time of the distant future.

One of the most challenging scientific endeavors of our time is predicting and understanding the future effects of the buildup of carbon dioxide and its effects on crops, polar ice, sea level, weather, and the global economy. Carbon is unique among elements in that it has such serious implications for our planet, and for our lives. Like it or not, the energy that drives the modern world, as it has since the first cavemen, is still largely derived from the chemical reaction of burning carbon compounds to produce carbon dioxide. It is the only element in the periodic table that has its own tax. Beginning with Finland in 1990, many countries now have some form of carbon tax as a way to stimulate migration to other forms of energy generation and reduce production of the greenhouse gas carbon dioxide. We will discuss some of these issues in chapter 8.

In our first biology class, we learn that life is based on carbon and that this element is unique in its ability to make strong bonds with itself and many other important elements. Yet the many roles of carbon and its unique properties and chemistry remain often underappreciated. Our main purpose in writing this book is to display as many of the glories of the sixth element as we can, from the earliest known writings and drawings to the latest nanotechnologies; from its birth in stars to its role in the formation of Earth to its many lives in the tools humans have created from it to sustain life and also to beautify and enhance it; to build; to invent; and to pass literature, art, music, laws, math, and other forms of accumulated knowledge to future generations.

In chapter 6, we focus on the amazing materials, tools, and technologies the sixth element has spawned, and some of the ways they have shaped history and our everyday lives. To take just one example, consider the important role carbon has played in making recorded history possible. As far back as twenty thousand years ago, charcoal was used in making the famous Paleolithic cave paintings in the Dordogne region of southwestern France, and for most people who are now alive, much of what they have learned was learned from reading letters printed in black carbon. Until recent times, carbon was used to create nearly all written or printed words. The Magna Carta, the Declaration of Independence, and, of course, all literature before computer-based word processing was written with microscopic carbon particles preserved in ink, or with pencil "lead," which is a mix of carbon and clay.

Another often unappreciated function of carbon in our lives is its fundamental role in providing us with color. Except for the ocean and sky, most of the color that enriches our daily lives involves carbon compounds, even if they are just a binder holding inorganic pigments together. Some color pigments are derived from coal tar or other petrochemicals. Our quite colorful, carbon-coated world provides a stunning contrast to Mars, the Moon, and Venus, our comparatively drab neighbors in space. Except for invisible carbon dioxide, these bodies do not contain appreciable amounts of carbon compounds. They are not covered with plants or paints, so they are mind-numbingly monochromatic, either gray or reddish.

The sixth element provides us with an astonishing number of capabilities that might be commonly overlooked. Steel, the backbone of most buildings, bridges, vehicles, and modern warfare, is not just iron; it is iron strengthened by the addition of small amounts of carbon that dramatically improve its properties. Cars, trucks, and buses ride on a miracle material, "rubber" tires, made of a mix of microscopic carbon particles held in a matrix of carbon polymers. The roads that these vehicles travel on are paved with either asphalt or concrete. Asphalt is a mix of petroleum and rock. Concrete is made with rocks, lime, and clay. Lime, a key ingredient, is made from roasted limestone. Limestone is by far the dominant form of carbon in our planet's outer layers. Just the production of concrete for roads and construction is responsible for an astounding 5 percent of the human-produced carbon dioxide that is being put into the atmosphere.

All of the foods that people and animals eat, as well as most of their packaging materials, are composed of carbon compounds. The diamonds you wear and those used to saw slabs of granite countertop out of cliffs are forms of pure carbon. The air you breathe is dominated by nitrogen and oxygen, but carbon dioxide, present in minor concentrations (0.04 percent), plays major roles in governing the long-term habitability of our planet and, of course, is the source of the carbon that allows giant trees and all plants to grow with energy provided by sunlight. This gas, although enjoyed in champagne and fizzy drinks, is commonly derided because of its role in global warming. It is an irony of nature that we can't live without this "toxic gas," because it is the "food of life" on our planet.

We will talk about plastics, carbon compounds that have revolutionized our society. There are natural plastics, such as amber, but humans have produced nearly 10 billion tons of synthetic plastics since the Second World War, usually from petroleum, of which carbon is the main component. Plastics have become ubiquitous both as litter and as products that we can't live without. Though plastics are often associated with waste and pollution of the Earth and oceans, they also enrich our lives in remarkable ways. The uses of plastic are seemingly endless, and some of our highest-tech materials are plastics. For example, a composite of epoxy and graphite fibers is used to make products that include spacecraft, tennis rackets, airplanes, skateboards, expensive cars, and warheads for ICBMs. The highest-quality displays for televisions and phones are made of organic light-emitting diodes (OLEDs). The use of the word "organic" in the OLED acronym does not mean that it was grown on a pesticide-free OLED farm, but rather that it is made of carbon-based molecules containing carbon-hydrogen and perhaps carbon-carbon bonds. This "misunderstanding" is an example of the common misuse or at least an alternative use of a scientific term. By scientific definition, CO, CO_2, and cyanide are not organic molecules, though almost every food item in a grocery store (including conventionally grown vegetables) is made of organic chemicals.

It is hard to imagine living in the modern world without plastics. A simple example is the elegance of a ziplock bag. For millennia, people used gourds, clay pots, baskets, or the internal organs of animals as containers to store precious food and water, but the utility of these containers pales in comparison to a strong, watertight bag made of polyethylene that is durable, transparent, thinner than a human hair, physically robust, nearly weightless, and can be used for years and years. However, we now use so many plastic bags, wraps, and containers that they have become a serious environmental nuisance.

We will see that, like many other things, the element carbon has both positive and negative potentials and attributes. The mining and use of coal and oil and even the inhalation of campfire soot has serious consequences on both present and past human health. The burning of fossil fuels has led to a buildup of carbon dioxide that is creating a frenzy of concern over human-induced global warming and sea level rise. Radioactive carbon-14 (^{14}C), which is made naturally by cosmic rays impacting nitrogen at the top of the atmosphere and also by nuclear bomb tests, provides a fantastic means to date events since the dawn of civilization, but, like coal, it also has side effects. Half of the radioactivity inside our bodies, a whopping four thousand disintegrations each second, is due to the decay of carbon-14. It is amazing to consider that a concentration of carbon atoms from our bodies can make a Geiger counter run off the scale, and that radioactive carbon made at the edge of the atmosphere as well as normal carbon are utilized in building the structure of plants that are ultimately eaten by us. The other half of our internal radioactivity comes from the decay of natural potassium in our bones. Although this radioactivity sounds alarmingly bad, the decay of carbon-14 in DNA has been proposed to play a possible role in genetic mutations that allow species to evolve over long time scales.

Our investigation of the sixth element will take us into these and other ethical aspects of its uses, and into the laboratories of great scientists—physicists, chemists, astronomers, biologists—who have contributed to our understanding of carbon and of what an element actually is, which played a crucial role in the history of science. It took scientists, or natural philosophers as they were once called, many centuries to figure out what an element is. At first, this could be done only by characterizing its behavior: if a particular substance always acted the same way in chemical reactions and could not be broken down into subordinate materials with different properties, then it was deemed to be an element. Only later was the link between element identity, atoms, and atomic structure understood.

The wondrous element carbon has truly shaped "our world" in the grandest sense that encompasses the origin and evolution of biology on Earth and extends to myriad nuclear and chemical processes that have and will occur over the entire spatial and time scale of the cosmos.

CHAPTER 1

The Discovery, Origin, and Dispersal of Carbon

Carbon has certainly been known about since the beginning of human history. The understanding of what carbon really is, however, did not occur until just over two centuries ago, when modern science was just emerging and instruments and experiments were being used to probe the nature of matter. At the time of the French Revolution, in the late 1700s, science in France was progressing as it was elsewhere in Europe. People were trying new things and thinking new thoughts. The solar system had recently been remodeled to a Sun-centered one; the laws of gravitation and motion were well in hand; astronomer William Herschel discovered our seventh planet; and chemistry was entering a period of revolutionary advancement as matter was recognized to be composed of elements that were indivisible and could not be decomposed into lesser components by chemical or mechanical processes.

French chemist Antoine Lavoisier (1743–94), with abundant assistance from his wife, Marie-Anne Paulze Lavoisier (1758–1836),[1.1, 1.2] played a fundamental role in what is often considered to be the birth of modern chemistry (figure 1.1).[1.3]

FIG. 1.1. The Lavoisiers. Antoine has long been considered the father of modern chemistry. In recognition of Marie-Anne Paulze Lavoisier's contributions as a woman scientist in the 1700s, some have suggested that she should be considered its mother. *Credit:* Jacques-Louis David, 1748–1825. *Antoine-Laurent Lavoisier (1743–1794) and His Wife, Marie-Anne Pierette Paulze (1758–1836).* 1788, oil on canvas. Metropolitan Museum of Art, New York.

The work was done with a methodical and highly scientific approach. Lavoisier carefully measured components before and after chemical experiments in order to quantitatively understand and document what had taken place. When gases were involved, they conducted the experiments in sealed chambers, so nothing could escape or intrude, and they measured the weight of the ingredients before and after the reaction took place. A major inference from this work was the Law of Conservation of Mass, referred to as Lavoisier's Law in France. *Mass is neither created nor destroyed in chemical reactions.* This is paraphrased in the sentence *"Rien ne se perd, rien ne se crée, tout se transforme"* (Nothing is lost, nothing is created, everything is transformed).

Like other chemists of the time, the Lavoisiers became interested in combustion. Many experimenters had focused on the combustion of various materials, including charcoal and graphite. Previous experiments showed that graphite and charcoal each produced "fixed air" (now known as carbon dioxide) when burned, gradually leading to the conclusion that these substances must be related. A complicating factor in interpreting these experiments was the widely held belief that combustion involved the release of a colorless, odorless, tasteless, and weightless substance called phlogiston. The interesting name comes from the Greek for "burned-up" or "flame." No one knew what phlogiston was or exactly how it interacted with other materials, but it was thought that when something burned, its phlogiston was released.[1.4] In England, Joseph Priestley had isolated oxygen from air and called it "dephlogisticated air." For a long time, no one realized that instead of being lost during combustion, something is added (oxygen).

The Lavoisiers conducted many experiments designed to find the nature of combustion and isolate the role of phlogiston

FIG 1.2. Giant burning glass of the Académie des Sciences, Paris, eighteenth century. Constructed under the direction of Antoine Lavoisier and others, it was used for chemical experiments. *Credit:* From *Les applications de la physique* by Amédée Guillemin (Paris, 1874). Oxford Science Archive/Heritage images/Science Photo Library.

(figure 1.2). As was their practice, they took great pains to weigh and measure the ingredients and the products. With this excellent scientific methodology, they found that whenever they burned something in a sealed container, the total weight of the burned substance plus the gas in the chamber always remained constant. In some cases, the material gained weight while the gases lost an equal amount of weight; in other cases, the burned substance partially or completely disappeared, with the gas gaining all of the weight that was lost by the sample. In every case, an ingredient of air, at the time known as "vital air" but now recognized (and named by Lavoisier) as the element oxygen, was taken from the air and combined with the material that burned. The experiments showed that combustion involved the addition, not the loss, of something to the material being burned. This discovery spelled the death of the phlogiston theory. In 1783, Lavoisier read his paper "Reflections on Phlogiston" to the French Academy of Sciences. This was an attack on the phlogiston theory of combustion. Phlogiston lost its mojo in the late 1700s and vanished into the dustbin of history.

Lavoisier's experiments demonstrated that during the combustion of charcoal, oxygen was taken from the air and "fixed air," or carbon dioxide, was produced. This happened whether the incinerated substance was charcoal or graphite. (Or diamond—Lavoisier actually vaporized a few diamonds in the quest to understand the chemistry of combustion! Diamonds totally burn up if heated in air above 750°C.) By the mid-1780s, the Lavoisiers and others had established that the flammable substance in graphite and charcoal, the material that combined with oxygen to make CO_2, was the same. The name "carbone" was applied, later "carbon" in English. The fact that carbon itself did not break down into other substances, and that it could be recovered after combining with something else, showed that carbon was an element. The Lavoisier experiments also showed that water was not an element, because it was divisible; it could be broken down into oxygen and hydrogen. At that time, no atomic theory had been developed, and apart from chemical behavior, what it meant to be an element was not established.

Antoine, unfortunately, did not live to find out. He was guillotined on May 8, 1794, along with twenty-seven other fellow members of the *Ferme générale,* an enterprise that paid the king for a contract that granted them the right to collect duties and taxes. He was a "tax farmer" and so was his wife's father, also guillotined on the same day. Tax farmers were among the most hated members of the old regime and were high on the hit list when the revolutionaries took charge, even when some of the profits were used to finance opening the door to a new era of understanding chemistry and the nature of matter. It was a social explosion, not a chemical one, that did in Antoine Lavoisier, and it is ironic that a person that created a revolution in science and named hydrogen, carbon, and oxygen would be intentionally killed during a period famously described as the best of times and the worst of times. But Marie-Anne Lavoisier was not among the seventeen thousand people guillotined during the French Revolution, and she organized the publication of her husband's final memoirs, *Mémoires de chimie*, a compilation of his papers and those of his colleagues demonstrating the principles of the new chemistry. She also rescued Lavoisier's notebooks and laboratory instruments, most of which are now curated at Cornell University.

In the centuries following the French Revolution and the discoveries of the Lavoisiers and many others who followed them, chemists and physicists developed a theory of atomic structure and an understanding of chemical elements that succeeded in explaining how atoms in gases and liquids interact with each other. In modern terms, a chemical element is characterized by its atomic structure, which consists of a nucleus containing protons (positively charged particles) and neutrons (no charge). Surrounding the nucleus at a great distance (relative to the size of the nucleus, though an infinitesimal distance by ordinary human standards) are electrons, which are constrained to orbit at fixed average distances from the nucleus (more on that in a moment).

The identities and chemical characteristics of each element are determined by the number of protons and electrons. Hydrogen is the simplest element, with just one proton in its nucleus and one electron; helium is next, with two protons and two neutrons as well as two electrons; then comes lithium, with three protons and three electrons and either three or four neutrons (the form having four neutrons is far more abundant in nature). After lithium come all the other elements, from beryllium (4 protons) to uranium (92 protons). There are elements beyond uranium, up to 118 protons.[1] More await discovery, although elements beyond uranium are expected to have short lives.

Carbon is a modest element with a nucleus containing six protons and six neutrons, usually surrounded by six electrons in its neutral state. What is so special about carbon? Why has it been singled out to star in this book all by itself? Naturally occurring elements can have almost a hundred protons, so if mass is a measure of elemental worthiness, carbon is at the low end of the stack. But despite its meager mass, in many ways carbon is among the most important elements, because without its special chemical and nuclear properties, life as we know it would not exist—and the Universe would be a very different place.

Why do we make this claim on behalf of an otherwise seemingly innocuous element, a collection of protons, neutrons, and electrons much like any other? Because a carbon atom has a very special talent for chemical interactions and because carbon's nucleus has a peculiar property with enormous consequences for the formation of other elements during the evolution of our Universe.

Chemical reactions involve the sharing of electrons between atoms, causing the atoms to stick together in the form of molecules. Atoms have built-in rules governing how many electrons are available to interact with neighboring atoms. The rules for carbon are liberal, so a carbon atom can combine chemically with many other elements, including other carbon atoms. You might say that carbon is very sociable, interacting with other elements freely.

The electrons orbiting an atomic nucleus are constrained by the laws of quantum mechanics to stay within certain orbital boundaries, sometimes called shells. The inner shell of any atom can contain only two electrons, while the next one can have up to eight electrons. Having just two electrons in total, helium's inner shell is filled and there are no openings or excess electrons to share with neighbors, which means it is very reluctant to react chemically with anything else. This is much like other elements whose electron shells are filled to capacity. Lithium, the third element, has its inner shell filled, but there is one lonely electron in the next shell, so lithium can react with other chemical species that can use a shared electron.

Carbon, with a total of six electrons, is filled up in the innermost shell, but only half of the eight openings in the next shell are occupied by its remaining four electrons. Carbon can share these four electrons with other atoms and can also accept up to four shared electrons provided by atomic neighbors. Thus, carbon has an extremely rich chemistry, being capable of merging with almost anything except the noble gases. The chemical reactivity of carbon explains why this element plays a role in combustion, as discovered by the Lavoisiers, and it explains why carbon is so ubiquitous in chemical and biological compounds here on Earth and elsewhere in the Galaxy.

Another factor in carbon's fame is its abundance. Carbon is a very common element in the Universe. Hydrogen and helium are by far the most abundant, with hydrogen atoms being about ten times more abundant than helium, and oxygen and carbon coming in third and fourth with roughly one thousand times fewer atoms than hydrogen. In a later chapter, we will consider why some elements are more abundant than others.

Despite its high cosmic abundance in the Universe, carbon is a relatively rare element in our planet. When the Sun and planets formed some 4.5 billion years ago, some elements, such as carbon and nitrogen, were unable to efficiently form or accrete solid matter. These elements remained in space as gas, and so, Earth, like its rocky neighbors Mercury, Venus, and Mars, has only trace amounts of these "uncondensed" elements. In the outer solar system, where temperatures were much colder, carbon, nitrogen, and other light elements could form solid materials that were used to form solar system bodies. Pluto, all of the outer planets, their icy moons, comets, and even asteroids have much higher abundances of carbon and nitrogen than the Earth and the inner planets.

Even though carbon is rare in the rocky inner planets, there was enough on Earth to lead to the formation of life and development of the widespread biological systems that we have today. Life arose on Earth from an inventory of carbon-based substances. The variety of organic compounds that can be made is nearly limitless. In a nutshell, this is why carbon is the element of life, and also why much of modern technology is based on carbon compounds. When the Lavoisiers isolated carbon and realized that it was a fundamental substance, a chemical element, they hit upon something far more important than anyone of their time could have realized.

We have described a few properties of carbon, and more details of its chemistry will follow in chapter 2. Now we will turn to the story of how carbon was created in the first place. The expansion of the Universe suggests one way. If the Universe is expanding, there must have been a time when everything was scrunched together. We can even figure out when that time was, by backtracking the expansion. Astronomers now think the start of the expansion, or the "Big Bang," was 13.8 billion years ago.[2]

Could carbon have been created during the Big Bang? George Gamow asked this question and started out to find the answer. Gamow, a flamboyant Russian émigré physicist, escaped the Soviet Union in 1933 and joined the faculty at Georgetown University, where he spent the next twenty-three years before leaving for the University of Colorado. (The University of Colorado treasures its association with Gamow, and a prominent building on the campus was named for him, the Gamow Tower, which can be seen for miles.) He was a pioneer in understanding nuclear fusion reactions, in which simple atoms (more precisely, their nuclei) merge to form heavier, more complex atoms. Gamow speculated that fusion must have occurred during the early stages of the Big Bang, when temperatures and densities were very high. He wondered whether today's abundances of all the elements could be explained by reactions that occurred during the Big Bang.

Gamow's graduate student Ralph Alpher modeled the formation of the elements during the time when the temperature of matter rapidly cooled from over a billion degrees.[1.5] In his PhD thesis, Alpher showed how all the elements could have formed, in the correct proportions as seen today, during the early stages of the Big Bang. This was recognized as a great discovery, and Alpher's doctoral thesis defense was attended by more than three hundred people. His central role in this great discovery was not immediately recognized, however, because of Gamow's preeminence in the field of stellar nuclear reactions.[3]

But could Big Bang nucleosynthesis explain the formation of carbon? Alpher thought so; he included carbon, along with all the other elements, in his theory. The modeling assumed the initial state was a dense sea of free neutrons that naturally decayed to protons and electrons. Neutrons and protons form deuterium and tritium, respectively the stable and the short-lived heavy isotopes of hydrogen. It was envisioned that most of the elements beyond the lightest ones would be formed by neutron capture reactions, where a nucleus takes on a neutron and then ejects a negative electron, a process called beta decay. After beta decay, the freshly taken neutron is gone and in its place is a proton, whose positive charge moves the nucleus to the next element in the periodic table. Further neutron captures followed by the emissions of electrons build up ever more massive nuclei until all the elements have been created, and of course, carbon, with its six protons and usually six neutrons, was part of the sequence. In their calculations, Alpher et al. got the right proportions of the elements as observed today. Now all of the elements in the Universe were accounted for, with 2 percent in heavier elements.

Everything was copacetic: the observed distribution of the elements was accounted for in the Alpher-Bethe-Gamow theory. End of story, right? Wrong. There was a huge problem. After the formation of helium, no further neutron-addition reactions were possible. No stable nuclei with five or eight nucleons (i.e., the total number of protons and neutrons) could exist. There was a break in the chain that prevented subsequent reactions from making heavy elements. (All right, a little lithium, with proton number 3, formed as well.) This entire sequence was over in less than four minutes after the Big Bang. After that, the Universe had cooled enough to prevent further reactions. By number, most of the atoms now in the Universe were made in the Big Bang, and the early Universe was an element desert devoid of elements heavier than lithium. The early Universe was no place for us or even planets like Earth—it was carbon free!

The Alpher-Bethe-Gamow theory of formation of all elements couldn't work. There was no reaction that could form elements beyond beryllium by simply adding more neutrons. This was a major defect in the theory, and that is where things stood until Fred Hoyle joined the fray, with his profound contributions to the understanding of how carbon formed inside of stars after the Big Bang. This was a critical step in seeing how the creation of carbon was a gateway to the formation of all heavier elements, nearly all the elements in the periodic table.

Hoyle, the only child of a rural family in Yorkshire, England, grew up a thoughtful, serious child. At an early age, he started to read books on astronomy and philosophy. At the same time, he had a rebellious side, which showed up when, for example, he concocted a scheme to miss school because it bored him. At an early age, Hoyle was a math genius and, upon reaching early manhood, he enrolled in college at Cambridge, earning degrees in both mathematics and physics. Already interested in astronomy, Hoyle focused on nuclear processes inside stars. And, having learned relativity theory, he became interested in cosmology, the origin of the Universe.[1.6]

Hoyle was well aware of the fact that elements with mass numbers greater than 5 or 8 couldn't have been formed during the Big Bang. But carbon's mass number, in its most common form, is 12: six protons and six neutrons. During a sabbatical stay at Caltech during the mid-1950s, Hoyle used his knowledge of nuclear physics to develop a model in which carbon was formed by the merger of three helium nuclei.[1.7, 1.8] He didn't yet know how this could happen; he just thought it must, because carbon is here—and we are here! This kind of reasoning is known as the anthropic principle; in other words, it is possible, in other Universes that might have existed, that no carbon was created, making for a dead Universe. But in this Universe, the fact that we are here means that somehow carbon was created.

And somehow the Universe had to have gone right through the gaps between atomic numbers 5 and 8. Realizing that adding three smaller nuclei together could make carbon, Hoyle envisioned a three-body reaction sequence, in which a beryllium

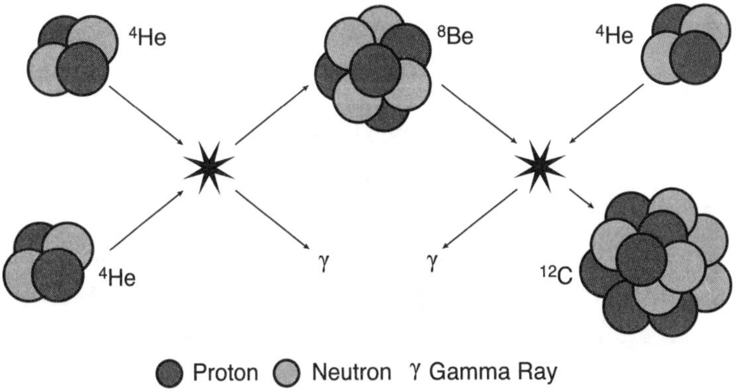

FIG 1.3. Triple alpha reaction where three helium nuclei fuse to form carbon.

nucleus is created by merging two helium nuclei to form ^{8}Be (that is, a mass 8 nucleus composed of four protons and four neutrons). The reaction goes like this:

$$^4He + {}^4He \rightarrow {}^8Be$$

There are two problems with this. One is that the reaction requires energy input; it's endothermic, meaning the reaction requires more energy input than it gives off. And, once formed, the ^{8}Be nucleus flies apart immediately. It only has a lifetime of 6.7×10^{-17} seconds, or a tenth of a billionth of a millionth of a second (see figure 1.3).

A third ^{4}He nucleus has to collide with the ^{8}Be particle before it disintegrates. Given enough particle speed, and if the density of particles is high enough, this reaction does occur, but it requires the nearly simultaneous impact of three particles. This keystone reaction is called the triple-alpha (triple-α) reaction. Energetic ^{4}He nuclei are called alpha particles. Now we have:

$$^4He + {}^4He \rightarrow {}^8Be + \gamma$$

$$^{8}\text{Be} + {}^{4}\text{He} \rightarrow {}^{12}\text{C} + \gamma$$

The γs are gamma rays, and ^{12}C is formed in an excited state. The second reaction, when a beryllium nucleus combines with the third helium nucleus, produces carbon with so much excess energy that it usually decays right back into three helium nuclei. Hoyle deduced that carbon must have a previously unknown excited state with an energy level that allows some of the carbon to linger until it reaches a stable state. The effect is called a resonance, and the "excited" period is now called the Hoyle state. But, even accounting for the Hoyle state making the reaction possible, fewer than one in a thousand reactions between Be and He produces a carbon that is able to release its excess energy and survive.

It seems that ^{12}C has an excited state at just the right energy level for a collision between it and a ^{8}Be nucleus to allow the two particles to stick together when they encounter each other, so the $^{8}\text{Be} + {}^{4}\text{He} \rightarrow {}^{12}\text{C} + \gamma$ can occur. Voilà! We have carbon, life is good, and the key stumbling block to making heavier elements is removed.

Based on the anthropic principle, Hoyle theorized that this excited state must exist; otherwise, we carbon-based life-forms wouldn't be here. But nuclear physicists at the time had trouble believing him, because no experiments had found the required excited state. Willie Fowler, Hoyle's colleague at Caltech, took up the challenge and did experiments aimed at finding the excited state of the carbon nucleus. And, lo and behold, Fowler found it, as Hoyle had predicted. The anthropic principle saved the day!

Hoyle's prediction was published in 1954, but almost nobody noticed. In a later paper, published in 1957, Hoyle and three others incorporated the triple-α process in their seminal paper in *Reviews of Modern Physics* describing element formation in stars. The four authors were Margaret and Geoffrey Burbidge, Willie Fowler, and Fred Hoyle (figure 1.4); the paper is now commonly cited as B²FH.[1.9] (In 2007, on the fiftieth anniversary of the publication, a major scientific conference was held in honor of the paper. Many notable researchers in the field of stellar nucleosynthesis paid homage to B²FH and reported their current research results.) Partly for that work, Fowler later (in 1983) received a Nobel Prize in physics. Hoyle was left out, even though he can be said to have made the most important contribution: discovering the missing link in bridging the N = 5 and 8 gap.[1.10]

But how does carbon, once made in the stars, get to Earth? Where is it now? The nuclear reactions needed to create carbon can occur only in very high density gas, in the centers of stars, and not anywhere near the surface. Yet somehow carbon gets out to interstellar space, and then to Earth. To explain how this happens, we will have to describe stellar evolution. For stars have lives, they evolve. They live and die, and some are reborn. If stars didn't evolve, carbon would remain on the inside of stars and not be available to play its many roles in our lives.

Stars form from collapsing interstellar clouds of gas and dust.[4] Our star, the Sun, will have a total lifetime of about 10 billion years, and it's halfway through, a middle-aged star. Stars come in many sizes and masses, and the reactions that create carbon, as well as the lifetime of a star, depend on both size and mass. It seems counterintuitive, but high-mass stars have short lifetimes, not long ones. That's because the nuclear reactions in stellar cores, where new elements are created, depend on temperature. The mass of the star determines the central temperature, its rate of reactions and evolution.[1.11]

Astronomers often use the mass of the Sun to describe the mass of other stars. A star with solar mass of 1 has the same mass

REVIEWS OF
MODERN PHYSICS

VOLUME 29, NUMBER 4   OCTOBER, 1957

Synthesis of the Elements in Stars[*]

E. MARGARET BURBIDGE, G. R. BURBIDGE, WILLIAM A. FOWLER, AND F. HOYLE

*Kellogg Radiation Laboratory, California Institute of Technology, and
Mount Wilson and Palomar Observatories, Carnegie Institution of Washington,
California Institute of Technology, Pasadena, California*

"It is the stars, The stars above us, govern our conditions";
(*King Lear*, Act IV, Scene 3)

but perhaps

"The fault, dear Brutus, is not in our stars, But in ourselves,"
(*Julius Caesar*, Act I, Scene 2)

FIG 1.4. Margaret Burbidge, Willie Fowler, Fred Hoyle, and Geoffrey Burbidge (B²FH) and their famous 1957 synthesis paper on the origin of the chemical elements. *Credits:* (*Top*) Courtesy the Margaret Burbidge Estate. (*Bottom*) "Synthesis of the Elements in Stars" by E. Margaret Burbidge, G. R. Burbidge, William A. Fowler, and F. Hoyle (*Rev. Mod. Phys.* 29, 547). © 1957 American Physical Society.

as our Sun. Stars with more than 25 solar masses generate nuclear energy so quickly that they only last for a few million years. In dramatic contrast, the lowest-mass stars have projected lifetimes of 10 trillion years, much longer than the present age of the Universe. Low-mass stars haven't had a chance to spread their carbon around; they keep it to themselves. In fact, stars with less mass than the Sun's mass don't create carbon in the first place, because their interiors never get hot enough. Once the hydrogen runs out, they just fade away.

Very massive stars die by blowing up very violently, exploding as supernovae. In the process, elements, including carbon, are spread around the Galaxy. Intermediate-mass stars don't explode as supernovae, but they eject matter in the form of smoke rings. These expanding clouds are called planetary nebulae because they superficially resemble planets when observed with small telescopes. William Herschel coined the term in the 1780s, and it has stuck.

Stars greater than 8 solar masses explode as core-collapse supernovae. Some leave ultra-dense neutron star cores, while others become black holes. Stars greater than 100 solar masses probably formed early in the Universe's history and exploded as "hypernovae" (we're running out of superlatives!), creating and emitting carbon in the process. In addition to production inside the cores of evolving stars, we now know that a significant fraction of the heaviest elements form during the explosive mergers of ultra-dense neutron stars. Such a merger was first observed in 2017 by detecting its production of a pulse of gravitational waves.

The intermediate stars, those between 0.5 and 8 solar masses, expand and undergo complex internal changes after they form helium and go through sequential stages of nuclear reactions that occur in the core. After the star is finished converting hydrogen to helium in its core, a shell just outside the core can start a hydrogen-burning process. The heat of these reactions gets bottled up, causing the outer atmosphere of the star to expand. That's where the terms "giant" or even "supergiant" come in, to describe a once ordinary star that expands by several tens or even hundreds of times its former diameter. When our Sun becomes a red giant, it will engulf Mercury, Venus, and probably Earth.

As the star expands, the core of the star continues heating until its carbon undergoes further reactions. At the same time, a shell of gas just outside the central core gets hot enough to start hydrogen burning, so a spherical shell of helium starts building up. This sequence goes on as a star creates several layers where different reactions occur. As helium is converted to carbon in the core, the layer just outside is changing hydrogen to helium; soon enough, the next level gets hot enough to make heavier elements. After helium, carbon, neon, and oxygen are created, then silicon, and so on. Eventually, for massive stars, its interior resembles an onion as the reactions eat their way outward (figure 1.5).

How many reaction stages a star goes through depends on its mass. The Sun will probably stop after carbon is created in its core. More massive stars will go through further reaction phases, creating oxygen, then neon, then magnesium, and eventually iron. This particular sequence of reaction comes from adding alpha particles; there are several other paths to make heavier elements that occur in massive stars.

In the red giant stage, the star's interior can be mixed up. Convection does this; heat flowing outward from the core region gets slowed on the way, causing the star's insides to roil, overturning the layers formed by successive nuclear reaction stages. In this way carbon gets to the surface. Because a giant or

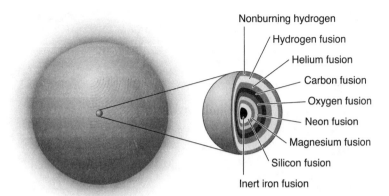

Nonburning hydrogen
Hydrogen fusion
Helium fusion
Carbon fusion
Oxygen fusion
Neon fusion
Magnesium fusion
Silicon fusion
Inert iron fusion

FIG 1.5. The "onion skin" model of the core of a giant star where the elements hydrogen to silicon are undergoing fusion reactions to form heavier elements.

a supergiant has very low surface gravity, its outer atmosphere is not bonded tightly to the star, and it can expand in a slow stellar wind. Carbon is expelled from the star in this wind. Giant stars are thought to be the primary source of carbon in the Galaxy.

Supernovae will violently end the lives of stars between about 8 and 50 solar masses. The reason: the chain of nuclear stages eventually ends with the formation of iron in the stellar core. Less massive stars never get that far before they stop making new elements. Iron stops the progression, because there are no more exothermic reactions. Iron starts sucking up energy from its surroundings rather than adding energy. Without a continuing source of energy, the core collapses inward. The outer layers follow suit more slowly, crashing into the core in a violent collision. The result is a huge outburst, a supernova that destroys the star. The expanding outer layers inject a variety of elements into the interstellar medium, carbon among them.

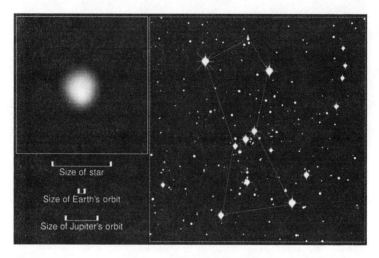

FIG 1.6. Betelgeuse, the spectacular red supergiant star in the Orion constellation. *Credit:* Andrea Dupree (Harvard-Smithsonian CfA), Ronald Gilliland (STScI), NASA and ESA.

Red supergiants and giants create most of the carbon in the Galaxy, with supernovae accounting for the small remaining percentage. Betelgeuse, the bright star in the shoulder of the constellation Orion (figure 1.6), is expected to explode in a supernova in the next hundred thousand years, or maybe it has already done it, and the light is on the way. It is more than 600 light-years away.

Stars of 0.8 to about 8 solar masses will become white dwarfs after their nuclear burning phases end. These stars are about the size of Earth, yet they have the mass of a star. Naturally, that means an incredibly high density, about 1 million tons per cubic meter. Our Sun will end as a carbon white dwarf, because its core nuclear reactions will end at carbon before heavier elements can be made.

In 2004, a white dwarf star whose core appears to be largely composed of crystalline diamond was discovered! This was inferred from its density and its pulsations, which tell astronomers how dense it is. Think of a ringing bell. A diamondlike structure is thought to be the only form of carbon that can exist under such conditions. The star's formal name is BPM 37093, but it is often referred to as Lucy, after the Beatles song "Lucy in the Sky with Diamonds." This amazing star is as massive as the Sun but smaller than Earth and spins thirty times a second.

We have considered what happens inside stars; what about the role of their elemental compositions? To understand how the Universe began, how the elements formed, and how the element carbon fits into the grand scheme of the cosmos, we'll need to call on knowledge of the compositions of stars.

Amazingly enough, the composition of the Sun and other stars has only been reasonably well understood for less than a century. In the past, it was often thought that the Sun was composed of heavy elements and similar to Earth. William Herschel, discoverer of the planet Uranus, is said to have stated that the Sun is "a cool, dark solid globe clothed in luxuriant vegetation and 'richly stored with inhabitants', protected by a heavy cloud-canopy from the intolerable glare of the upper luminous region."

It wasn't until after 1925 that a good case was made that the Earth and Sun have drastically different compositions and that common stars have compositions rather similar to that of the Sun. In 1802 and 1814, William Wollaston and Joseph von Fraunhofer independently discovered that the Sun's light, dispersed into different spectral colors with a prism, has dark lines, now called Fraunhofer lines. Nearly a half century later, it was shown that these lines are caused by light being absorbed at specific wavelengths by specific elements in the outer layers of the Sun. The strongest Fraunhofer lines are due to hydrogen, iron, sodium, and calcium. In the beginning of the twentieth century, it was commonly argued that the solar Fraunhofer lines were consistent with the belief that the Sun and the Earth have similar elemental compositions. Stars of different colors have quite different spectra, and the prominence of different lines was used to classify them into the lettered groups O, B, A, F, G, K, and M. It was thought that the spectra of these stars differed according to their different compositions.

The revolution in understanding the composition of the Sun and other stars was due to the work of Cecilia Payne (later Cecilia Payne-Gaposchkin) (figure 1.7). She solved the problem, though she had to overcome many obstacles, such as being a woman of science at that time in history and weathering the disbelief of others, including Henry Norris Russell, an eminent astronomer of the day. Her doctoral thesis is now seen as one of the biggest breakthroughs in astronomy. Payne was English born, reared, and educated—until she tried to enter Cambridge University for a doctorate, but, being a woman, didn't get in despite her stellar performance at lower-level schools. American astronomer Harlow Shapley, director of the Harvard College Observatory, awarded her a position in a new doctoral program for women. This was an era when women, generally underappreciated at the time, made historically profound discoveries at the observatory.[1.12, 1.13]

Payne's 1925 thesis explored the underlying physics of stars to explain their spectra and temperatures. Using the work of Meghnad Saha, an Indian physicist who developed a theory of the ionization of gases, she found that the temperature and the density of the outer layers of a star dictate which elements dominate its spectrum. For the Sun and most stars, she found that fifteen elements between lithium and barium had similar relative abundances to what is found in the Earth. She also found that the spectral differences between stars was largely due to their temperatures, not their compositions. Using Saha's theory, she determined the temperature of stars by the states of ionization

FIG 1.7. The stellar cast of the play *Observatory Pinafore*, performed by staff and students at the Harvard College Observatory on the last day of 1929. From left to right are Peter Millman, Cecilia Payne, Henrietta Swope, Mildred Shapley, Helen Sawyer, Sylvia Mussels, Adelaide Ames, and Leon Campbell. *Credit:* AIP Emilio Segrè Visual Archives, Shapley Collection.

of their elements. Most of her thesis findings were respected; however, her work also had the astounding finding that hydrogen and helium are the main elements in the Sun. This was radically different from what "was known" about stars. When Russell reviewed a draft of her thesis, he thought that the implied abundance of hydrogen was nonsense—"It is clearly impossible . . . that hydrogen should be a million times more abundant than the metals." When Payne published the results of her work in the *Publication of the National Academy of Sciences*, she chose to note that her "improbably high abundances of hydrogen and helium" were "almost certainly not real." But it was real, and this finding fundamentally changed our understanding of the Sun and stars. Hydrogen and helium are actually 98 percent of the Sun's total mass, and, in this regard, the Sun is dramatically different from our rocky planet, which is dominated by magnesium, silicon, iron, and oxygen atoms. The Sun's carbon content is 0.04 percent by mass, and it contains ten times more carbon atoms than silicon, magnesium, and iron atoms. Carbon, in spite of all of its importance to us, is only a trace element in Earth and a minor element in the Sun. Her results indicated that the mass of the Universe was mostly in hydrogen, the lightest element and one that we now know was formed, along with helium, in the first few minutes of the Big Bang. The highly controversial part of her thesis opened the door to understanding how stars generate energy, how the Universe began, and how most of the elements around us were made in a chain of processes that started with hydrogen as the initial feedstock. There was initial resistance to Payne-Gaposchkin's revolutionary conclusion, but in time, it was fully accepted. Astronomer Otto Struve described her work as "the most brilliant PhD thesis ever written in astronomy."[5]

We have seen how detailed investigation into arcane ideas such as the fictitious phlogiston led to the initial understanding of carbon and its identification as a fundamental element. We also looked at how the sixth element formed, how its abundance compares with other elements, and some aspects of its chemistry. In the next chapter we will consider why carbon is such a unique element because of its natural flexibility to form compounds with wide-ranging properties.

CHAPTER 2

The Chemistry of Carbon: Why Is It So Special?

Scientific investigation of mysterious materials and processes led the Lavoisiers to discover and name carbon, which many consider to be the birth of modern chemistry. It is fitting that the sixth element, with its incredible ability to form compounds, played a groundbreaking role in our understanding of nature. Its remarkable chemistry is what really sets the sixth element apart from all of the other elements.

Wake up and smell the:

If you don't recognize this, it is a diagram of a caffeine molecule, aka 1,3,7-Trimethylxanthine. This molecule does not actually have an odor, but it is in the coffee that billions of people drink (and smell) every morning. Eight carbon atoms bond with hydrogen, oxygen, and nitrogen to form a perky, addictive molecule. This is just one example of carbon's importance in chemistry—the science of atoms held together by chemical bonds—and in our lives.

In years past, before the diversions of the internet and heightened concerns with child safety, many kids had chemistry sets that allowed them to mix liquids and powders of various sorts and get new colors, bubbles, smoke, flames, the ability to harden liquids into solids, and even modest explosions. If the kids followed directions, the risk of harm was minimal, and many were inspired with a wonder for the marvels of chemistry and science in general. The more sophisticated chemistry sets, usually for high school students, provided a valuable introduction to chemistry as a science and also sparked a fascination with carbon and its many compounds.[1]

When you imagine a chemist, you might think of a mad scientist in a lab, wearing a white coat and misshapen wire-rimmed eyeglasses. This gives the impression that chemistry is arcane, depraved, and something of a danger to us all. The words "toxic" and "chemical" are often linked together in some people's minds, and the internet contains numerous suggestions on how to avoid chemicals in our lives. Just search on the web for "chemical free" and you'll be amazed at the number of products that are promoted as being free of chemicals. (The misuse or redefining of chemical terms is not uncommon; for example, almost no one presently alive has ever used foil actually made of tin or a pencil with lead in it.)

These days, some people casually consider chemicals and chemistry to exist mainly in the lab or to be synonymous with pollution. Not so! Chemistry is the science of atoms or molecules sticking together or breaking up, getting or giving. Breaking a molecule usually requires energy input, and forming a molecule usually emits energy. Chemicals are all around us (not to mention in us) all the time. The air we breathe, the water (or other liquids) we drink, your car and its fuel, the trees and plants all around us are of course made of chemicals. Some are natural; some are human-made.

Before the seventeenth century, there was hardly a difference between chemistry and alchemy. A fair analogy is astrology versus astronomy: one is based on fable or belief; the other is a science that is objective and testable. The gradual evolution of alchemy to chemistry began with Islamic alchemists.[2.1, 2.2] The science of chemistry developed a robust framework of understanding why substances react with each other. Reactions are both testable and predictive. One famous alchemist was Isaac Newton. While discovering the principles of calculus, mechanics, light, gravity, and telescopes, he dabbled in alchemy, without much success.

A number of scientists tried ways of grouping elements by their properties. In 1869 Russian scientist Dmitri Mendeleev invented a chart ordering elements by their atomic weights.[2.3, 2.4] The chart also groups element chemical affinities and actually predicted the existence of several elements that were unknown at the time the chart was first made. A modern version of the chart, the periodic table (figure 2.1), hangs on the walls of most chemistry classrooms.

The number above the element's abbreviation is the number of protons and neutrons, and the number below is the number of protons, also called the atomic number. Isotopes of an element differ in the number of neutrons in their centers. All isotopes of a given element have the same number of protons in their nucleus and generally have very similar chemical properties. Carbon has

Key

| relative atomic mass |
| **atomic symbol** |
| name |
| atomic (proton) number |

1	2											3	4	5	6	7	0
																	4 **He** helium 2
7 **Li** lithium 3	9 **Be** beryllium 4											11 **B** boron 5	12 **C** carbon 6	14 **N** nitrogen 7	16 **O** oxygen 8	19 **F** fluorine 9	20 **Ne** neon 10
23 **Na** sodium 11	24 **Mg** magnesium 12											27 **Al** aluminium 13	28 **Si** silicon 14	31 **P** phosphorus 15	32 **S** sulfur 16	35.5 **Cl** chlorine 17	40 **Ar** argon 18
39 **K** potassium 19	40 **Ca** calcium 20	45 **Sc** scandium 21	48 **Ti** titanium 22	51 **V** vanadium 23	52 **Cr** chromium 24	55 **Mn** manganese 25	56 **Fe** iron 26	59 **Co** cobalt 27	59 **Ni** nickel 28	63.5 **Cu** copper 29	65 **Zn** zinc 30	70 **Ga** gallium 31	73 **Ge** germanium 32	75 **As** arsenic 33	79 **Se** selenium 34	80 **Br** bromine 35	84 **Kr** krypton 36
85 **Rb** rubidium 37	88 **Sr** strontium 38	89 **Y** yttrium 39	91 **Zr** zirconium 40	93 **Nb** niobium 41	96 **Mo** molybdenum 42	[98] **Tc** technetium 43	101 **Ru** ruthenium 44	103 **Rh** rhodium 45	106 **Pd** palladium 46	108 **Ag** silver 47	112 **Cd** cadmium 48	115 **In** indium 49	119 **Sn** tin 50	122 **Sb** antimony 51	128 **Te** tellurium 52	127 **I** iodine 53	131 **Xe** xenon 54
133 **Cs** caesium 55	137 **Ba** barium 56	139 **La*** lanthanum 57	178 **Hf** hafnium 72	181 **Ta** tantalum 73	184 **W** tungsten 74	186 **Re** rhenium 75	190 **Os** osmium 76	192 **Ir** iridium 77	195 **Pt** platinum 78	197 **Au** gold 79	201 **Hg** mercury 80	204 **Tl** thallium 81	207 **Pb** lead 82	209 **Bi** bismuth 83	[209] **Po** polonium 84	[210] **At** astatine 85	[222] **Rn** radon 86
[223] **Fr** francium 87	[226] **Ra** radium 88	[227] **Ac*** actinium 89	[261] **Rf** rutherfordium 104	[262] **Db** dubnium 105	[266] **Sg** seaborgium 106	[264] **Bh** bohrium 107	[277] **Hs** hassium 108	[268] **Mt** meitnerium 109	[271] **Ds** darmstadtium 110	[272] **Rg** roentgenium 111							

1 **H** hydrogen 1

Elements with atomic numbers 112 – 116 have been reported but not fully authenticated

* The Lanthanides (atomic numbers 58 – 71) and the Actinides (atomic numbers 90 – 103) have been omitted.

Relative atomic masses for **Cu** and **Cl** have not been rounded to the nearest whole number.

FIG 2.1. The periodic table, a common sight in every chemistry classroom. *Credit:* © 2023 International Union of Pure and Applied Chemistry.

two stable isotopes, ^{12}C and ^{13}C, as well as ^{14}C, which is radio-active. Carbon-13 and -14 are quite rare, so that the averaged mass of carbon on Earth is 12.011. Pure carbon-12 is defined as having an atomic mass of 12, and it is the standard by which the other elements' masses are measured.

We have referred to carbon's importance to us: this element dominates the chemistry of all living things, animals and plants alike; some of the most useful materials in our lives are made of carbon-based stuff. No carbon; no Earth; no people. But why? To understand why carbon and all of the other elements behave chemically as they do, we have to take an excursion into physics on an atomic scale. Wolfgang Pauli was a Nobel Prize winner from Austria and part of a group of theoretical physicists along with Max Planck and Niels Bohr. He came to the US, where he spent the rest of his career and developed the laws of quantum mechanics, a set of very odd rules that successfully describe the nature and behavior of matter at atomic and even smaller scales, and a field of study that pushed scientific inquiry into a new era and opened a window on our Universe, from atomic and mo-lecular bonds to the structure of white dwarfs and neutron stars.[2.5, 2.6]

An essential foundation of the theory was that no two elec-trons can occupy the exact same quantum state. Why? Pauli explained this as a result of the "exclusion principle," developed in 1925, one of the foundations of modern physics. Pauli and other physicists, in trying to explain why atoms exist, asked, Shouldn't an electron fall into the nucleus as energy is lost when the atom emits light? Certainly, this occurs on a macroscopic scale: a moon orbiting a planet falls closer to the planet if the moon loses energy for some reason. This much we knew from Newton's laws, but they didn't explain atoms and molecules. On the quantum scale, things are different. Rather than a gradual change in an electron's position relative to the nucleus, discrete jumps between levels occur, emitting or absorbing a photon in the process. When an electron drops between levels, a photon of light is emitted, and when a photon is absorbed, an electron goes from one level to a higher one.

Nobel laureate Richard Feynman, one of the smartest people of his time, said, "If you think you understand quantum mechanics, you don't understand quantum mechanics," and "I think I can safely say that nobody understands quantum mechanics." But, while some aspects of quantum mechanics seem very alien to previously established principles and our everyday experiences of the natural world, it turned out to be a quantitative and quite predictive way to understand much of what goes on in chemistry and atomic physics in the realm of the very small.

As a consequence of the Pauli exclusion principle, we now understand that only two electrons can be in the innermost orbit;[2] eight can be in the next state; eighteen in the next; and so on. Hydrogen, the simplest element, with only one proton in its nucleus, usually has its single electron in the lowest energy level. Carbon, with its six electrons, has two electrons in the lowest level and four in the next higher state that can be involved in chemical bonds. These odd rules of nature give carbon the superlative chemical properties that enabled the complex processes that led to the formation and long-term evolution of life. At least on our planet, no other element has played such a role.

As discussed in chapter 1, an atom with a filled shell has a hard time interacting with other atoms. That's why the "noble gases," such as helium, neon, and argon, usually can't bond with other atoms. Helium has two electrons, completely filling up the first shell. A helium atom just flies around alone, perfectly content as is, or it is physically trapped inside solid material.

The same resistance to chemical bonding applies to other noble atoms with full shells such as argon, neon, and krypton. These closed shell elements could never plausibly provide a pathway for the processes needed for life.

Carbon is just the opposite from a noble gas in its ability to form bonds. The first level is filled by its two electrons, and the second level has four out of eight possible spaces. With only four electrons in a possible space for eight, it leaves the second level with four unfilled electron states. Therefore, carbon can add up to four electrons or give away as many as four, providing many ways that a carbon atom can bond with another atom or molecule.

Among common elements, only silicon has a similar chemistry in that it may add electrons to its outermost level, which has four (out of a possible eighteen) in that level. Like carbon, a silicon atom has four electrons, called valence electrons, that are capable of forming bonds with as many as four other atoms. Silicon is not a cosmically rare element (it is the seventh most abundant in the Galaxy)—but carbon is about four times more abundant. On Earth, silicon is much more abundant (26 percent by mass) than carbon (only a few hundred parts per million), because carbon was rare in the rocky materials that formed Earth.

It is intriguing that carbon is so rare in our planet. The sixth element was abundant in the early solar system, and under the right conditions it could form millions of compounds. In spite of all its wonders, in the region of space where Mars, Earth, Venus, and the Moon formed, it was largely stranded in gaseous nebular molecules such as carbon monoxide and could not efficiently form solid building materials of the types that made our planet.

In our carbon-biased view, most people believe that carbon is essential for life that is based on chemical processes and is even remotely analogous to what we have on Earth. The extraordinary properties of this atom are related to the way that atoms stick together to form molecules; they have to form bonds. There are two main types of bonds in general use by atoms to form molecules.[3] [2.7] The less common of the two is called ionic bonding. (Our celebrity element, carbon, almost never forms compounds by ionic bonding.) This is when two atoms encounter each other, and one atom would like to add an electron (in other words, the outermost shell has a vacancy), so an electron jumps across the space between atoms. That leaves one atom with negative charge, and the donor atom, now missing an electron, has positive charge. Opposite charges attract each other, resulting in a strong electrostatic bond between the two atoms.

A classic example of ionic bonding is salt. Table salt is made of cubic arrays of just two elements, sodium and chlorine (figure 2.2). In this case chlorine is the electron recipient, and sodium is the electron donor. Now each atom has a filled outer shell, and each is content as long as they are bonded to each other.

In chemical notation, this reaction can be represented by an equation:

$$Na^+ + Cl^- \rightarrow NaCl = salt$$

The elements Na and Cl stand for sodium and chlorine. The plus and minus signs indicate the charge.

The other common kind of bond between atoms, and the one that carbon uses, is covalent bonding, where atoms share electrons in molecules.[2.7] The term "covalent" refers to the sharing of electron orbits; we can think of it as the overlapping of the outermost (or "valence") electron orbits. An electron pair belongs to adjoining atoms, in the sense that they orbit nuclei close together. More than one pair can orbit, forming stronger covalent bonds as the number of paired orbits grows (figure 2.3).

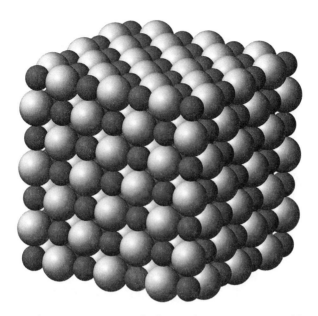

FIG 2.2. Salt atomic structure. The large spheres are negative chlorine ions, and the smaller dark ones are positively charged sodium ions. *Credit:* Benjah-bmm27 via Wikimedia Commons.

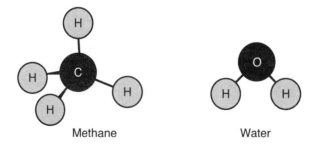

FIG 2.3. Methane and water.

In some cases, two electron pairs bond two atoms in place; this is called a double bond. As this involves four electrons, it is naturally stronger than a single bond. An example of a double bond is in ethylene, where two carbon atoms are joined by a double covalent bond between two carbon atoms (figure 2.4).

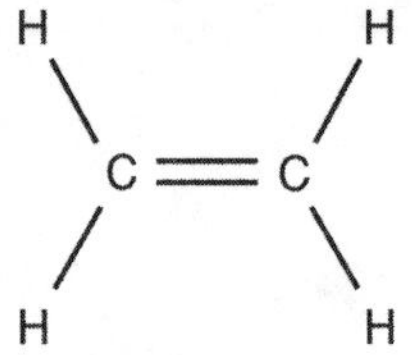

FIG 2.4. Ethelene (C_2H_4) with a double bond.

Even triple bonds are formed in some molecules. An example of a molecule with a triple bond is the very poisonous acid hydrogen cyanide (HCN), which has a triple bond between carbon and nitrogen, and a single bond between carbon and hydrogen. In chemical notation, it looks like figure 2.5: the three bars indicate the triple bond, while the single bar represents a single bond:

FIG 2.5. Hydrogen cyanide (HCN) with a triple bond.

Carbon normally forms covalent bonds with other atoms. In rare cases, like calcium carbide, carbon does form ionic bonds. There are variations in covalent bonds that strongly determine physical properties. There are countless possibilities for single, double, and triple bonds. For example, the simplest amino acid, glycine, NH_2CH_2COOH, has one double bond and six single bonds. The structure of a glycine molecule looks like figure 2.6:

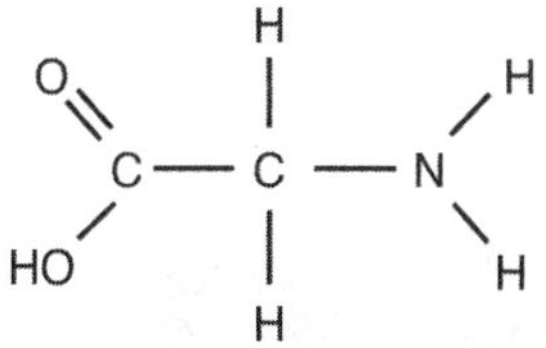

FIG 2.6. Glycine (NH_2CH_2COOH).

Many people were taught that pure elemental carbon comes in just three forms: coal, graphite, and diamond. Now we realize that is only part of the story. Carbon's chemistry gives it many arrangements (allotropes) of atoms; the atoms stick to each other in different configurations. Not all pure carbon forms are common or even natural—as we have seen, there are many forms of pure carbon, ranging from single atoms to diamond.

In the progression from simple atoms to complex configurations are carbon chains. As the name suggests, these are rows of carbon atoms, joined together in a line. There can be two atoms or many atoms. Carbon chains, like single atoms, are rare on Earth, but can exist in space. In interstellar space the simplest carbon molecule, C_2, can be lengthened to form C_3, and even longer chains are observed. The champion is a chain of eleven carbon atoms with a hydrogen bound to one end and a nitrogen bound to the other end. Even longer carbon chains are possible but haven't been observed yet.

Graphene comes next. Graphene consists of hexagonal rings of carbon atoms joined together at the edges.[2.8] Graphene essentially has only two dimensions. It is a sheet one atom thick (about 0.34 nm) that can, in principle, be nearly infinite in lateral extension. Think of chicken wire, and you get the picture (figure 2.7).

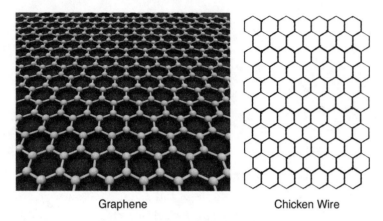

Graphene   Chicken Wire

FIG 2.7. Graphene and chicken wire.

The first to find a method of fabricating graphene were Andre Geim and Kostya Novoselov, who shared the Nobel Prize in Physics in 2010. Geim in particular is a funny kind of guy who won the "Ig Nobel Prize"[4] in 1997 for using a strong magnetic field to levitate a frog! In a more serious vein, Geim and Novoselov pulled thin flakes off graphite using tape. This works because the forces that hold graphite sheets together are very weak, similar to those in mica, a mineral that is easily split into thin sheets. Graphene has many potentially very useful properties. It has the highest strength-to-weight ratio ever seen, and is the strongest substance known. A gram of graphene can cover a football field, support the weight of a one-pound mass, and yet be invisible because of its thinness.

But that's not all. Electrons can move around freely in the sheet, meaning it has high electrical conductivity. Graphene also readily conducts heat along the plane of carbon atoms. The properties of graphene are highly dependent on direction and are considered to be remarkably anisotropic—that is, its properties vary with the direction that they are measured from.

Graphene has numerous possible uses for future electronic components and other devices.

Graphite is made of stacked planes of graphene. It is geologically formed at high temperature and pressure. When mined, it is shiny with a metallic luster, and it was originally called plumbago (from *plumbum*, the Latin word for lead) because of its resemblance to galena, a sulfide mineral mined for lead since Roman times. Graphite is found in every continent, and it is a quite robust form of carbon. Given enough time and heat, even diamonds will degrade into graphite. The bonding of carbon atoms within a sheet of graphene is extremely strong, but the planes are very weakly held together. Thus, graphite is slippery, as the graphene planes slide across each other.

Sheets of carbon can curl to form three-dimensional structures. The efforts of UK chemist Harry Kroto to understand the origin of spectral features seen in interstellar space led to the famous discovery of the spherical carbon molecule C_{60}.[2.9] Kroto, Robert Curl, and Richard Smalley vaporized graphite with lasers and examined the masses of the clusters of carbon atoms that formed.[2.10] They saw a distinct peak at the mass of sixty carbon atoms that stood above lesser and larger clusters of carbon atoms. Mysterious at first, it was soon realized that the molecule was a closed structure composed of hexagons and pentagons made from sixty carbon atoms with both single and double bonds. This is the same geometry as a soccer ball, and the amazing molecule was named buckminsterfullerene, for Buckminster Fuller,[5] or buckyball for short (figure 2.8).

In 1996, Kroto, Curl, and Smalley won the Nobel Prize in Chemistry for discovering fullerenes that include the spherical C_{60} and the elliptical C_{70}. (And Kroto, being an Englishman, was knighted by Queen Elizabeth in 2000.) Buckyballs were an exciting new discovery, but only tiny amounts were made in the

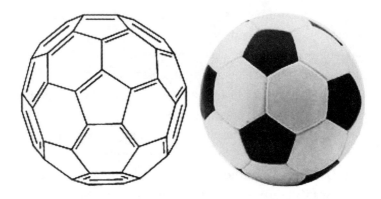

FIG 2.8. Buckyball (C_{60}) and a soccer ball. *Credits: (Left)* Pumbaa80 via Wikimedia Commons (CC BY-SA 2.5). (*Right*) 痛 via Wikimedia Commons (CC BY-SA 4.0).

experiments, and they were somewhat of a laboratory curiosity until Wolfgang Krätschmer and Donald Huffman discovered a way to make large numbers of buckyballs easily and inexpensively.[2.11] They used electric arcs to vaporize graphite into a low-pressure inert gas and deposit the resulting particles on quartz slides. They were trying to duplicate the shape of an ultraviolet spectral feature with a particular wavelength seen in the interstellar medium. It was suspected that the feature, seen everywhere in the sky by space telescopes, might be due to the presence of buckyballs. Under the right conditions, they could produce the astronomical signature with the thin films of small particles on their slides. They noted that the condensed material could be dissolved in benzene to produce a beautiful purple liquid. This was a simple way to make and purify buckyballs, separating them from other experimental products. When dried, the solution formed spectacular crystals composed of tiny spheres, each containing sixty carbon atoms. Thus, the investigations of the optical properties of carbon-bearing dust in our Galaxy led to fundamental discoveries of a new form of carbon and its efficient production for industrial utilization.

There are also intermediate states of pure carbon between sheets and buckyballs. Presently, the most promising are carbon nanotubes that have single or multiple walls. They can readily be made, but it is not clear if they play any significant role in nature or even if ideal single-wall nanotubes are produced in nature. Carbon nanotubes have great potential for revolutionary applications in medical fields and electronics. The blackest material known is made of carbon nanotubes because of their incredible ability to absorb light. Light goes in, but it doesn't get out.

There is also amorphous carbon or glassy carbon that has no long range order—that is, no regular pattern that repeats itself throughout the crystalline structure—but even amorphous carbon is not totally amorphous, as it usually does show some ordering at the scale of clusters of small numbers of atoms. An additional case of the common misuse of scientific terms is that "crystal" glassware is not made of crystal but of amorphous material.

Most of the carbon that we deal with is in molecules, where carbon is bonded to other atoms. Carbon-bearing molecules can range from the very simple, like carbon monoxide (CO) with two atoms, to the most massive stable molecules that have been synthesized, such as PG_5 composed of 20 million atoms.

An important class of carbon-bearing molecules are the polycyclic aromatic hydrocarbons (PAHs), such as naphthalene. (The term "aromatic" refers to its smell; many PAHs emit odors, starting with naphthalene, also known as mothballs.) As previously noted, carbon atoms like to stick together because of the four electrons and the four vacancies, which, when bonded, fill an eight-electron shell. In PAHs, there are usually alternate single and double bonds, and hydrogen atoms surround the carbon structure, as in naphthalene, $C_{10}H_8$ (figure 2.9).

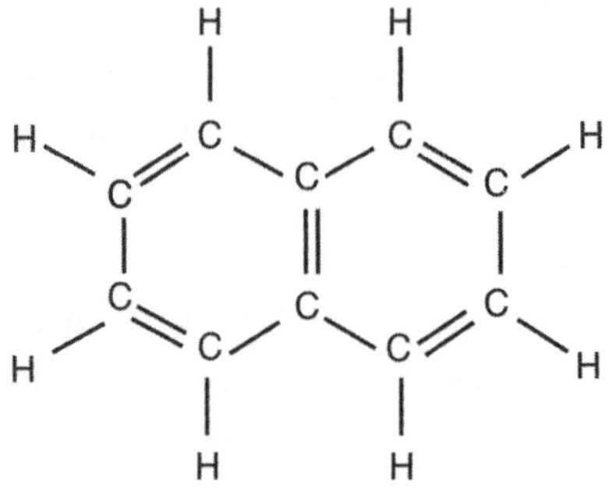

FIG 2.9. Naphthalene $(C_{10}H_8)$.

PAHs are everywhere, whether natural or synthetic, and they exist in space as well as our daily lives. Important sources of PAHs include the atmosphere, wood-burning stoves, volcanic gas, forest fires, gasoline, paint, heaters, soot, furnaces, coal, cigarette smoke, asphalt, pollution, either in the air or in lakes and oceans—the list goes on. Most are cancerous, a danger to breathe in. Food broiled to a high temperature creates PAHs, and the air that all humans breathe carries dust particles that contain PAHs.

Another important class of carbon molecules is the vast array biomolecules that are made and utilized by life. The four major types of biomolecules are carbohydrates, lipids, nucleic acids, and proteins, including deoxyribonucleic acid, or DNA. We will cover DNA with additional detail in chapter 4, but for now we want to highlight that this key molecule consists of two strands, interconnected by weakly held hydrogen atoms. The strands twist around each other, forming a double helix. James Watson and Francis Crick, with critical contributions by Rosalind Franklin, found this structure in 1950. The discovery of DNA was one of the most important scientific discoveries ever made. (In honor of Franklin, a European rover designed to operate on Mars has been named *Rosalind Franklin*.) Amino acid

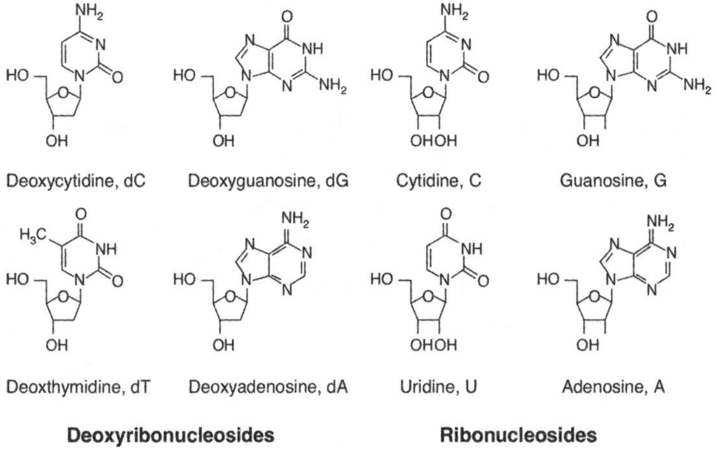

Deoxycytidine, dC Deoxyguanosine, dG Cytidine, C Guanosine, G

Deoxthymidine, dT Deoxyadenosine, dA Uridine, U Adenosine, A

Deoxyribonucleosides **Ribonucleosides**

FIG 2.10. Amino acids.

molecules are the building blocks of proteins (figure 2.10). Large protein molecules have many vital functions in life, including providing structure for cells and organisms, transporting molecules from one location to another, catalyzing metabolic reactions, replicating DNA, responding to stimuli, and making photosynthesis possible, the ultimate source of energy that drives terrestrial biology. Although there are hundreds of different amino acids, only twenty are utilized by life to assemble proteins, long chains of the residues of amino acid molecules. The largest human protein is titin, used as a spring in muscle. It can contain over thirty thousand amino acid molecules. The most abundant protein on Earth is rubisco, made from up to about five hundred amino acid molecules. Rubisco plays a crucial role in the complex magic of photosynthesis that makes biomass from atmospheric CO_2 and makes life possible on our planet. Because of the number of components and the presence of side chains, proteins are fantastically complex molecules. Some evolved and were fine-tuned for their tasks over geologically long time scales. The code for sequencing chains of amino acids into complex proteins is carried in DNA.

Carbon is just one element in the vast periodic table of elements, but, as we have previously discussed, it is quite unique due its extraordinary ability to form bonds. With its unique chemical dexterity, it stands out as much more than just being the only "black as coal" element in its most common pure form. Critically important is its ability to make complex long chain molecules and also make side branches. It can form single, double, and triple bonds, not only with itself but also with a number of other elements. The chemical prowess of this element is truly a miracle of nature.

CHAPTER 3

Carbon on Earth and in the Solar System

As humans, we have an inherent interest in our origins and history. And an essential part of the story of the many complex pathways that ultimately led to humans is how the carbon in us and our planet got here. We saw in chapter 1 that the interstellar carbon atoms that ended up inside Earth were involved in a series of intriguing processes. Astronomers and other scientists deduce the origin and history of our planet by studying Earth, meteorites, and other planets, and by observing young stars that are surrounded by disks of planet-forming gas and dust. Ancient stargazers found part of the story, by noticing that planets follow similar paths crossing the sky, implying that they formed from a disk, but that's about all, because they thought Earth was the center of the Universe. Most of their models of the solar system were geocentric.

Two millennia ago, Greek philosopher Aristarchus of Samos suggested that the Earth circled the Sun, but the true watershed of this idea did not begin to be widely considered until 1543, when the Polish astronomer Nicolaus Copernicus published *On the Revolutions of the Heavenly Spheres*, a description of a

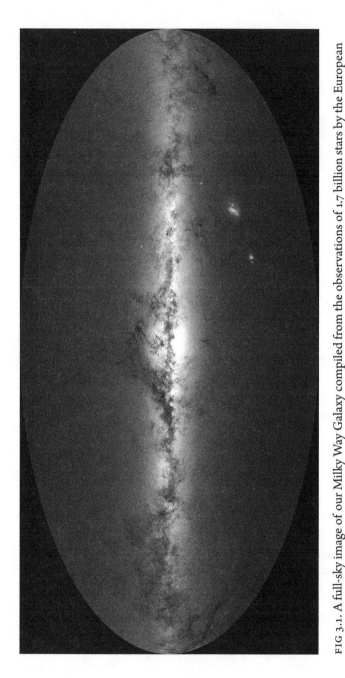

FIG 3.1. A full-sky image of our Milky Way Galaxy compiled from the observations of 1.7 billion stars by the European Space Agency's Gaia space telescope. The center of the Milky Way is in the image center; the two bright regions below center on the right are the nearby Small and Large Magellanic Cloud galaxies; and the dark regions are due to dust blocking starlight in the plane of the Milky Way. This band of carbon-rich dust can easily be seen with the naked eye. *Credit:* ESA/Gaia/DPAC (CC BY-SA 3.0 IGO).

Sun-centered Universe. He based his theory on the motions of the planets, especially Venus, whose apparent brightness changes as it goes around the Sun. Incidentally, the book was officially banned by the Catholic church in 1616. In the early 1600s, German astronomer Johannes Kepler[1] derived three important laws of the movements and places of the planets, centered around the Sun, and finally, in the late 1600s, Isaac Newton figured out the physics behind planetary motion.

Then the ultimate question was, how did the solar system of bodies orbiting the Sun form? And when? First we'll tackle the question of when.[3.1] The age of the Earth was highly controversial in the nineteenth century. Beyond biblical arguments, the key scientific debate was whether the Earth's age was measured in millions or billions of years. Physicists, geologists, and even biologists and paleontologists got into the act. There was an undercurrent of distrust or at least disdain among the disciplines, adherents of each one thinking that only they understood the whole picture.

Based on the time it would have taken for the Earth to cool from an initial molten state to its current temperature, William Thomson (aka Lord Kelvin), a British physicist, in the mid-1800s calculated an age of 24 to 400 million years. He later refined his estimate to 100 million years, and even later whittled it down to 24 million years. In his calculations, however, Thomson was unaware of important factors that, ultimately, vastly increased estimates of the age of Earth: one was convection, a heat-exchange process that increased the rate of heat transfer in the planet's interior; another was the ongoing source of heat in Earth's interior in the form of radioactive decay of uranium, potassium, and thorium.

At the same time, geologists were convinced that the Earth was very much older, with an age in the billions of years—or perhaps infinite. The idea that Earth's age is infinite was put forward by James Hutton and later Charles Lyell, based on the "uniformitarianism" concept, that everything affecting the Earth was cyclic, and therefore the Earth didn't necessarily have a beginning or an end. A major part of their argument was the slow process of sedimentation, which formed the layers in the Earth's crust, and the counterbalancing effect of erosion, destroying those layers at the same time, an interchange that could go on for a very long time.

Biologists now chimed in, saying that the evolution of species, from early primitive life to modern complexity, required much more time than mere millions of years, or even hundreds of millions of years. Observed rates of sedimentation, combined with the progression of complexity of fossils with increasing depth, showed that the evolution of plant and animal species occurred over a very long time.

The final nail in the coffin of a young Earth was the measurement of age by radioactive decay times of certain elements. Radioactive decay occurs because some elements are unstable— that is, they spontaneously emit subatomic particles such as electrons, positrons, or alpha particles, which are helium nuclei containing two protons and two neutrons. A few elements are so unstable that they can decay into fission fragments more massive than helium nuclei. Perhaps the best known radioactive elements are uranium, plutonium, and the rare form of carbon, carbon-14 (^{14}C). A radioactive element has a half-life,[2] the time needed for half of that element to decay. For ^{14}C, the half-life is 5,730 years, short enough to be useful for age-dating human artifacts. We will discuss the interesting history of carbon-14 in chapter 6.

The half-life of carbon-14 is much too short for age-dating the Earth. Uranium is an excellent measuring tool; its most abundant isotope, ^{238}U, has a half-life of 4.5 Gyr (billion years). After a long chain of decay steps, ^{238}U and ^{235}U ultimately end up as stable lead isotopes ^{206}Pb and ^{207}Pb. Results from uranium dating give an irrefutable age of 4.4 Gyr for the oldest minerals that have survived our planet's history of destructive processes, and so the half-life of uranium closely matches the age of the Earth that registers its history. By including measurements from meteorites, the formation age of our planet is pushed back to 4.55 Gyr.

The first definitive age of Earth and the solar system was determined by Clair Patterson at Caltech in the 1950s.[3.2] In his work measuring lead isotopes, Patterson discovered an enormous environmental background of lead contamination that had to be dealt with. The lead background is not "natural" but has accumulated from millennia of human uses, ranging from Roman plumbing to paint used to cover houses. A major source was lead tetraethyl, which was added to gasoline to make engines run more smoothly. Somewhat like the greenhouse gas CO_2 quietly accumulating in our atmosphere from fossil fuels, human-processed lead had been accumulating in the global environment, and no one had noticed or cared. The environmental background of lead was found to cause significant health problems such as lower IQ in children. Against considerable opposition from industry, Patterson's heroic efforts in determining the age of our planet eventually resulted in banning the use of lead in products such as gasoline, plumbing, water coolers, paint, dishes, and toys, a classic example of basic science changing the world that we live in.

Earlier we mentioned meteorites being important in pushing back estimates of age. The oldest dated materials that formed in the solar system are found in meteorites. They date to 4.567 billion years, and this is widely assumed to also be the age of the Sun. After the Big Bang, about 13.8 billion years ago, over 9 billion years passed before interstellar matter gravitationally collapsed to form the Sun and its planets, moons, asteroids, and comets.

Now to the question of Earth's formation. The solar system's planets formed from a disk of gas and dust that orbited the young Sun at the time of its birth. Using rapidly improving methods in optical, infrared, and radio telescopes, we can observe and study disks of gas, dust, and small rocks around young stars. Observations show that such disks do not retain major amounts of gases, including H_2O and CO, after a few million years. So processes that require gases must have occurred very quickly.

In the case of our Sun, the disk that the planets formed from is often referred to as the solar nebula. The nebula was composed of material gathered together by gravitational forces, and initially it was dominated by hydrogen, which has always been the most abundant element in the Universe. The solar-like composition (by mass fraction) of the original preplanetary material was approximately 73 percent hydrogen and 25 percent helium, with all the other elements making up the remaining 2 percent of the mass. Of the 2 percent, the carbon abundance was right behind oxygen and ahead of nitrogen. The number of carbon atoms in the material just barely exceeded the total number of magnesium, silicon, and iron atoms, the atoms that, along with oxygen, dominate the mass of our planet.

The disk of gas and solid particles orbiting the young Sun had a natural tendency to clump, which led to the formation of larger and larger bodies. The processes involved were complex and are still not well understood, although there has been considerable recent observational and theoretical progress in this area. It is likely that many different processes were involved in making planets from the disks, and their relative importance varied in location and timing. It is also likely that their importance varies from star to star, both in systematic and predictable ways that may depend on the mass or composition of the star and in random ways that are stochastic (depend on chance). The diverse properties of planets detected around other stars are strong evidence that planet formation, while common, is not the result of simple and predictable processes. Complicating our understanding of planet formation is definitive evidence that planets can migrate from their birth orbits to both larger and smaller orbits. This can happen when the disks still retain gas, and even billions of years after the gas was gone.

For the solar system, the formation of solid bodies began with condensation, the formation of solids from cooling gas.[3.3] Chemical elements that could form solids were initially in tiny solid interstellar grains when they were part of the ensemble of interstellar materials that formed the Sun and its disk, but nearly all of the pre-solar grains were vaporized in hot regions of the disk. Then condensation could occur in disk regions cool enough for the solids to form. For water ice, the gas must be cooler than −150°C, and for iron metal it must be cooler than about 1,000°C. The gas in the inner regions of the disk, near the Sun, was too hot for ice, but robust rocky materials and iron metal could exist as solids. Very close to the Sun, where the temperature exceeded 2,000°C, it was so hot that no solids could exist, and so it was a dust-free zone. But the outer regions of the disk, near Pluto, were cold enough that water and even supervolatiles such as carbon monoxide, methane, and nitrogen could condense.

When tiny solids formed in the disk, they immediately began to form larger and larger bodies. There are numerous growth pathways, but one of the first processes is simply through collisions, where particles stick and snowball to such large size that they can fall and concentrate toward the disk midplane. The traditional view is that bodies of kilometer size and larger formed by a long chain of collisions and sticking, but growth by collisions can be inhibited when collision speeds become high enough to break up bodies, and when bodies reach meter size, the friction from gas headwinds causes their orbits to decay in sunward spirals. The traditional collisional scheme also has difficulties explaining how planets assembled in the short time scale that they appear to have formed in.

The formation of planets is now widely believed to have had considerable assistance from gas processes that led to concentrations of small, solid particles. One category of these growth ideas involves "streaming instabilities"; these are dynamic gas effects in disks where gas causes dust to concentrate and form clumps. In a gas disk spinning around a star, even a local region with an unusually high density can cause small rocks to migrate toward it and increase the local mass of solids. If there is a gas clump in a disk, gas tailwind forces on particles orbiting just interior to the clump cause them to move outward, while particles exterior to the clump move inward due to headwinds. If the density of solids becomes high enough, the self-gravity of a clump containing dust and boulders can quickly and gently collapse to form large solid bodies.

With time, growth increases until the solids in the disk become dominated by planetesimals, or preplanetary bodies, with diameters up to several thousand kilometers. The final assembly of planets like Earth involves high-speed violent collisions of bodies as large as our Moon and Mars. The formation of our Moon is widely thought to be the result of a collision of a Mars-size body with Earth during the later stages of its growth.

The locations where materials accumulate influence the composition of the bodies they form. Even among the asteroids between Mars and Jupiter, there is a trend in which carbon- and water-bearing minerals increase with solar distance. Generally, bodies that form closer to the Sun have lesser amounts of more volatile compounds, including those that contain carbon and nitrogen. This is the case even though planets and smaller bodies can migrate. It is quite ironic that Earth, the only body in the Universe known to contain life, was formed with only minuscule amounts of carbon, nitrogen, and water. The planets beyond Mars contain higher amounts of carbon, but, with the exception of Pluto and its clan, they have no surfaces. The outer regions of the giant planets Jupiter, Saturn, Uranus, and Neptune are just gas. Though the atmospheres may contain tiny particles of solids or liquids, there is no planetary surface.

Long after its formation, the inventory of solar system bodies looked like this: the planets and their moons, plus an assortment of lesser bodies that we now consider to be asteroids, comets, or other bodies orbiting the Sun. The lesser bodies that survived never accreted into planets, and they survived both destruction or ejection from the solar orbit for billions of years because they resided in rare gravitational refugia far from massive planets.

The planets can be grouped by their general physical properties. All of the planets between Mars and Pluto are massive; they do not have surfaces and are called giant planets. Uranus and Neptune are called ice giants, not because they are composed of ice but because ices were an abundant component of the rocky/icy solids that they formed from. The accumulation of ices and solid organics caused the "ice giants" to have a higher fraction of carbon than the other two giant planets. Jupiter and Saturn are called "gas giants" because they formed from gases and, like the Sun, their masses are dominated by hydrogen and helium, elements that were largely in gas in the early solar system. Because of their great mass, the giant planets contain most of the solar system's carbon that is not in the Sun but, like in the Sun, their carbon is highly diluted by more abundant lighter elements.

The planets interior to Jupiter are called terrestrial planets because they are relatively Earth-like and, most important, they have surfaces. All of the planets contain carbon in their interiors, on their surfaces, and in their atmospheres, but the terrestrial planets are of special interest because of their potential to provide surface or near-surface environments suitable for life, even if carbon is actually rare inside. As previously mentioned, it is rather amazing that carbon, the fourth-most-abundant element in the Sun, makes up only a few hundred parts per million of Earth's mass. Some terrestrial planets orbiting other stars could have formed in ways by which they accumulated much greater fractions of carbon-rich solids, and thus could have total carbon abundances a hundred times that of Earth. Studies of exoplanets orbiting other stars indicate that there is considerable diversity among terrestrial planets and the processes that formed them. Around other stars, processes have made some terrestrial planets much larger than any solar system terrestrial planet and much closer to their stars than Mercury, the closest planet to our Sun.

The most carbon-rich bodies in the solar system are not the Sun or the planets, but smaller bodies such as comets and asteroids, leftover planetary building blocks that have escaped collisions with planets or ejection from solar orbits for more than 4 billion years. Most of these bodies are smaller than one thousand kilometers, they are charcoal black, and most orbit in special regions of space where they can escape drastic gravitational perturbation by planets. If they ever do go close to planets, they are often perturbed to paths that either escape the solar system or impact the Sun or a planet.

These small bodies include the main belt asteroids between Mars and Jupiter and the comets that spend most of their lives in two regions beyond Neptune. One source region for comets is the Kuiper Belt just beyond Neptune, and the other is the Oort cloud, which extends to distances more than a thousand times the size of Neptune's orbit. The scientist who "found" this cloud of comets was Jan Oort, a Dutch astronomer. He didn't actually observe any of the bodies but theorized them by analyzing comet orbits and deducing that they fall inward from a huge spherical distribution of many millions of bodies. And the Kuiper Belt is a flattened, ring-shaped, distribution of orbiting bodies whose existence was suggested by Dutch-born astronomer Gerald Kuiper long before its existence was discovered.

Most of the asteroids and probably all of the comets are carbon rich because they formed in cold, distant regions. In addition to rocky and organic materials, comets contain large amounts of frozen water, carbon monoxide, and other volatiles. Many of the asteroids also retain impressive contents of organic materials, and many also originally had appreciable levels of water ice when they formed. In asteroids too close to the Sun, the ice either evaporated or melted to form water that could chemically react with solids to form hydrous minerals such as talc or serpentine. When the solar system was young, it contained a vastly greater number of carbon-rich asteroid- and comet-like bodies, most of which either became part of planets as they formed or were ejected on escape orbits from the Sun.

Besides planets, asteroids, and comets, there are two additional classes of bodies that can retain carbon for the billions of years of solar system history: the moons and the dwarf planets. Moons orbiting planets are phenomenally diverse in their properties and carbon content. Some, such as our Moon and Io, the most volcanically active body known, contain only trace amounts of carbon. Others that formed and were preserved at low temperature contain major amounts of carbon. One of the most impressive cases of a carbon-rich moon is Titan, Saturn's largest moon. Titan has a frigid nitrogen-rich atmosphere with a higher surface pressure than our own. Titan's density suggests that the moon's interior is close to a 50:50 mix of water and rock. Its atmosphere contains nearly 3 percent methane. Other hydrocarbons form in the atmosphere and sometimes rain on the surface. The surface has peculiar hydrocarbon lakes that appear to change over time and can reflect sunlight like mirrors.

The five known "dwarf planets" all formed far from the Sun and contain abundant carbon. The first dwarf planet, the large spherical asteroid Ceres, was discovered in 1801. The second and largest is Pluto, discovered at Lowell Observatory in 1930 by Clyde Tombaugh. The upgrading of the largest asteroid and the downgrading of the famous planet Pluto into the newly defined dwarf planet category was a controversial decision made, by vote, at a 2006 meeting of the International Astronomical Union. Ceres, Pluto, and Pluto's large moon Charon are the only bodies in this class that have been studied up close by spacecraft. As seen by the *New Horizons* spacecraft in 2015, Pluto has a spectacular surface with previously unimagined activity. The surface is largely frozen nitrogen, with smaller amounts of water ice, frozen carbon monoxide, and methane ice. There is evidence of dramatic geologic activity, including a large region that looks like convection cells and dramatic mountains composed of water ice. Pluto's interior is mostly rocky material, probably with a good amount of carbon, and there is a very thick mantle of water ice (figure 3.2).[3]

Our closest planet neighbors are vastly different from Earth, and they have had quite unstable histories compared to our generally stable, water-covered planet. Little is known about

FIG 3.2. A view of the spectacular planet Pluto as seen by the *New Horizon* spacecraft. The smooth region is young and covered with nitrogen ice. The mountains on the left are made of water ice. Some of Pluto's mountains have coatings of methane frost. *Credit:* NASA/ Johns Hopkins University Applied Physics Laboratory/Southwest Research Institute.

carbon in the interiors of Mars and Venus, but their amazing CO_2-dominated atmospheres are reasonably well studied. The surface pressure on Mars is less than 1 percent of ours, and its surface is frigid due its distance from the Sun and only feeble greenhouse warming in its atmosphere. The Venusian atmosphere is one hundred times denser than ours and has such extreme greenhouse heating that the surface temperature is hot enough to melt lead. Earth has at least as much carbon as Venus has in its massive atmosphere, but ours is locked up in limestone below the surface and other regions deeper within the planet.

How did Earth and the other terrestrial planets get their carbon?[3.4] Earth formed from the short-lived disk of material that orbited the young Sun. Originally, our planet's carbon atoms arrived in what would become the solar system as part of the ensemble of interstellar gas and dust that the Sun and planets formed from. The first carriers included gases such as carbon monoxide and methane, plus larger, more complex molecules, as well as a wide range of carbon-bearing solid materials that included organic materials, ices, silicon carbide, and even individual carbon atoms that were energetically implanted into otherwise carbon-free solids. It is possible that complex organic molecules that escaped destruction during the processes that assembled the solar system could have played direct roles in the prebiotic chemical evolution that led to the formation of life. In this sense, complex organics older than the Sun might have "kick-started" the evolutionary path that led to life. It is also possible that all or nearly all of these molecules could have been destroyed in the thermal and radiation environments that are involved in making stars and planets.

We have mentioned that the original planet-forming disk was hot in its inner regions and cold enough in its outer regions that ices could condense to form solids. In the warm region where Earth and the other terrestrial planets formed, ices could not exist as small grains. This region was too warm to condense carbon compounds, and there was no straightforward way for carbon solids to form from nebular gas and be abundantly incorporated into planets. In the cold regions of the solar nebula, carbon could condense as ices that could be partly transformed to organic molecules by radiation and other processes. Most of Earth's mass arrived as solids, but carbon was largely stuck in gas in the inner solar system and could not form solids. Accordingly, our planet, as well as the other terrestrial planets, is carbon poor.

Earth accreted from solids as small as dust to chunks as large as Mars (half the size of Earth). Some of these impacting bodies probably formed locally, while some came from more distant locales in the solar system, including regions where carbon- and water-bearing silicates were much more abundant than in our volatile-poor planet. In the solar nebula there was a distance from the Sun beyond which conditions were cold enough for water ice to condense on grains. This is called the snow line. It is likely that much of Earth's water and carbon came from materials that formed in cold regions beyond the snowline. Some materials in this region contained ice, and some contained materials such as hydrated silicate minerals that formed inside kilometer-size planetesimals by chemical alteration of rock in contact with water from melted ice. This can only happen inside bodies that become warm and large enough to have interior pressure and temperature sufficiently high to melt ice and retain water. Hydrated silicates are important carriers of water because they can retain "bound water" at much higher temperatures than ice can. Ice can only be retained inside small rocky bodies in space if the temperature is below about −130°C, but hydrated silicates can hold water up to temperatures of 400°C where these minerals break down and release bound water.

Earth is still accreting carbon-rich fragments of asteroids and comets, materials that formed beyond the snowline. In a typical year, the infalling matter is dominated by interplanetary dust at a rate of forty thousand tons a year. Most of the particles are smaller than the thickness of a human hair and their organic carbon content is usually around 10 percent—vastly higher than the carbon abundance in terrestrial planets.

In meteorites, cosmic dust samples, and samples collected directly from comets and asteroids, we see that many early solar system materials are mixes of materials that were transported over large distances and formed at different times. The bulk of Earth's mass was accreted when Earth had grown to become a massive body, and incoming solids fell to Earth at speeds of many kilometers a second. Their kinetic energy exceeded the binding energy of their molecules, and most existing compounds in them were destroyed by violent heating during impact.

Except for meteorites and cosmic dust that came later, all of the original carbon that was delivered to our planet was processed by extreme conditions that occurred on the early Earth. This included such high temperatures that the entire planet was covered by a deep magma ocean of molten rock. These processes broke chemical bonds in carbon compounds and transformed them. Carbon atoms deep inside our planet had an iron-loving (siderophile) chemical character that probably caused a major fraction of the planet's carbon to dissolve into the metallic iron that sank to the center of the Earth to form our large iron core. This very same process also depleted Earth's crust and mantle of other iron-loving elements such as gold, platinum, and nickel, and they became concentrated in the core. It seems paradoxical that a carbon-poor planet, the only proven life-bearing body in the Universe, probably has most of its treasured carbon totally isolated from the surface, where it could play any role in life processes. It is locked up in the planet's hellish center, where the temperature is over 5,000°C (as hot as the surface of the Sun) and the pressure is over 3 million atmospheres.[3.5, 3.6]

Because carbon is so visibly apparent in all the life-forms and biomass on our life-covered planet, it is sometimes imagined that our carbon is mainly near Earth's surface, but we have seen that this is not the case: most lies hidden within the planet. But Earth has a layered structure, and carbon is present at all levels from the top of the atmosphere to the core. Below the surface, some of the large-scale layering is related to differentiation, a fundamental process that occurred early and was common in

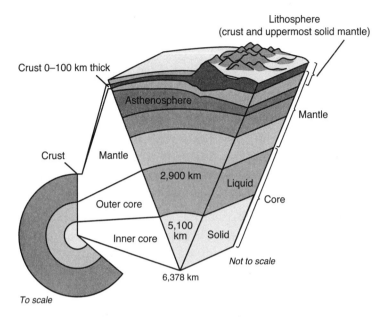

FIG 3.3. Earth interior. *Credit:* USGS.

terrestrial planets, large moons, and even the first-formed asteroids. For Earth, differentiation formed the mantle, with the interior core made of metallic iron enriched in iron-loving elements. The core formed because metallic iron and silicates do not mix in their molten or plastic states, and the metal, being much denser than silicate materials, sank toward the center of the Earth. Earth's overall density is about 5.5 grams per cubic centimeter, the crust is about 3, and the core averages about 12.2. The inner core is solid due to the high pressure, but most of the iron core is molten (figure 3.3).

The interior structure of Earth can't be directly probed from the surface; the Kola Superdeep Borehole, in Russia, is 12.3 km deep, whereas the radius of the Earth is 6,373 km. That's only a scratch. Some other means have to be found, and there is one that offers instruction, though we can't exactly plan them: earthquakes. When the ground is shaking, geologists, by the timing and placement of the quakes, can deduce the structure of several layers of interior Earth, from the rigid lithosphere to the center.

In our exploration of carbon, we have discussed its discovery, origin, and chemistry, and have shown quite a few interesting things about the distribution of carbon on Earth. In chapter 4, we will discuss carbon's role in life on Earth and elsewhere. In chapter 5, we will explore issues of carbon in our whole Galaxy and those related to life on exoplanets orbiting other stars. The first detection of carbon dioxide in the atmosphere of an exoplanet was made by the James Webb Space telescope.[3.7] What can we generally expect for exoplanets? As in our solar system, we expect that other planets will also contain carbon, but in differing amounts and forms. Some is likely to involved in life, but many planets are likely to be lifeless either because they don't provide adequate environments or life just never formed.

CHAPTER 4

Carbon and Life on Earth and Elsewhere

Apart from carbon's role in the grand evolution of the Universe, the sixth element is particularly interesting to us humans because of its central role in biology. Life has been here for most of our planet's existence, but evidence of the earliest life become less clear and more controversial as we probe deeper into the past, because we live on such an active planet, and older materials are continually altered or destroyed. It is generally agreed that the first reasonably convincing evidence for life goes back to the period between 3.5 and 4 billion years ago. The earliest time is of special interest, because the time before 3.9 billion years ago includes the period in which the Moon was struck by numerous large impactors that severely cratered its oldest surfaces. Although there is some controversy about what happened, it is commonly called the Late Heavy Bombardment (LHB), and it was responsible for forming most of the circular features that you can see on the face of the Moon with binoculars.[4.1] Assuming Earth was exposed to similar impacts at that time, it has been estimated that it would have received about forty events that would create 1,000 km diameter or larger craters or impact basins. Life evolved early in history, but the geological record is sketchy before the LHB. We may never know if life predated this time.

Life did form, and it survived the profound environmental changes that have occurred over the next few billion years. Because of its unique chemical dexterity, the sixth element forms the foundation of the incredible molecular machinery that is the basis of life, at least life "as we know it." How life is distinguished from nonlife is a common topic of discussion, from grade-schoolers to the loftiest pundits of the broad science of astrobiology. Leapfrogging over complexities of life definitions are the general ideas that life eats, excretes, reproduces, and evolves. A more formal definition of the minimum requirements is that it has the ability to reproduce and has the capability to undergo gradual change (i.e., evolve) as its habitat changes.

How and when did Earth's abundant life form? The most abundant elements in living organisms are carbon, hydrogen, oxygen, and nitrogen. With the exception of helium, they are also the most abundant elements in the Sun and the Universe. This group of four light elements is often referred to by the acronym CHON. From the beginning, long before Earth had free oxygen in its atmosphere, carbon was present at the onset of life's formation. Carbon constitutes about 18 percent of our body weight. Hydrogen, which dominates the composition of the Universe, is only about 10 percent of the human body by weight, and mostly it is also bound in water molecules. Any atmospheric hydrogen gas present at the time of Earth's formation has long since escaped to space. Due to its light weight, free hydrogen travels at such a high speed at the top of the atmosphere that it readily escapes to space. Our present atmosphere maintains small levels of hydrogen due to a balance between sources such as volcanoes and losses at the top of the atmosphere. Oxygen (in the form of water) is the most abundant element in people, at about 65 percent. Even in typical rocks, over 90 percent of the volume is occupied by oxygen atoms. Considered by volume, our planet is nearly pure oxygen! Nitrogen dominates our atmosphere but is only a trace element in the bulk Earth, yet its biological and geological cycles are an essential element for life because of its role in amino acids and genetic material.

Carbon, hydrogen, and oxygen were named by Antoine Lavoisier, and all four of these critical elements were discovered in a short period of time, part of the rapid advance of scientific activity and thought. Carbon was known since ancient times, but it came to be recognized as an element due to the work of Henry Cavendish, Carl Wilhelm Scheele, and Joseph Priestley, and was officially listed as an element in 1789 by Lavoisier. Hydrogen was discovered by Cavendish, an English physicist, in 1766, but it had been produced earlier and not known as an element. Cavendish also was the first to note that when burned, it produced water. Oxygen was discovered independently by two or three chemists in different countries. The first was in 1772 by Scheele in Sweden, but he did not publish his findings until after it was discovered and published in 1774 in England by Priestley. Priestley called it dephlogisticated air, which is how it was known until Lavoisier later named it oxygen. Lavoisier claimed to have independently discovered oxygen in 1775 in France, but there is controversy over how independent the discovery was. Nitrogen was discovered in 1772 by Scottish scientist Daniel Rutherford, although Scheele, Cavendish, Priestley, and others were also working on it. Rutherford discovered nitrogen as the residue of air where oxygen had been removed by burning.

The cosmically abundant CHON elements are critically needed for life. One of our planet's greatest mysteries is how this bunch of light elements led to the formation of life. For centuries, people believed that our carbon-based life could simply arise spontaneously from nonliving matter—for example, that a nutrient broth would spontaneously grow mold and bacteria. In support of this idea was the observation that maggots can appear in a pile of garbage without any intervention from outside. But, by the mid-1800s, Louis Pasteur had demonstrated that isolation of sterile, nonliving matter prevented new life-forms from appearing within it. This finding led to the germ theory: preexisting microorganisms present around us in the air lead to contamination, and life does not appear spontaneously. As for the garbage observation, flies did it. Of course, a pile of garbage is nirvana to them, so much so that they lay their eggs in it. A few days later, little babies (maggots) appear. They eat the garbage, the parents having long ago left the scene. So, the maggots didn't appear from nothing.

An alternative to the spontaneous formation of life was proposed in 1907 by Swedish chemist Svante Arrhenius, who suggested that life on Earth was introduced billions of years ago from space, arriving originally in the form of microscopic spores that float through the Galaxy, landing here and there to act as seeds for new biological systems. This idea was called the panspermia hypothesis. Several arguments make it seem unlikely, though. Such spores would take a very long time to permeate the Galaxy, and, most important, it seems very unlikely that the spores could survive the hazards of long-term space travel, where they would be exposed to cosmic rays, X-rays, ultraviolet light, shocks from stellar explosions, and high-speed collisions with interstellar dust. Besides, this only displaces the argument in any case: life has to form somewhere, and then it has to disperse its spores into interstellar space. If not on Earth, then where?

The panspermia hypothesis, a form of exogenesis for the origin of life on Earth, has a few modern-day advocates, but only a few. Again, we find Fred Hoyle at the center of things. His advocacy of the steady-state Universe makes no demands on the creation of life; it was always here. But, finally accepting an expanding Universe, he had to find a way to explain a constant matter density in spite of the expansion, which would lower the density if no refills were available. Therefore, Hoyle invoked a cosmic refilling station: new matter forming in intergalactic space to counteract the expansion. To be consistent, Hoyle thought life appears at all times and all places, forever and everywhere. The "spore" theory (exogenesis) would explain this.

Hoyle and his followers argued that organic materials, including bacteria, could exist on interstellar dust grains, and that the introduction of these grains onto the Earth's surface not only brought life to our planet in the first place but even today causes earthly epidemics of flu and other illnesses. Hoyle's science fiction novel *The Black Cloud* had a plot involving germs invading Earth. Hoyle's longtime colleague, Chandra Wickramasinghe, has kept this fantastical idea alive with the suggestion that the COVID-19 virus that produced a global pandemic was carried by a meteorite that exploded in our atmosphere in 2019 in the vicinity of Wuhan, China.

Other arguments for exogenesis are that there has not been enough time in Earth's history for random collections of atoms to form the complex molecules of life, or that there was no viable mechanism on the early Earth to produce large molecules such as RNA and DNA. This argument has been countered by the observation that certain arrangements of molecules are heavily favored, and that the combinations that lead to life are not random but are determined by the affinities of certain molecules for each other. Some have suggested that materials such

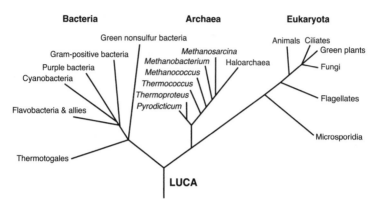

FIG. 4.1. This phylogenetic tree, derived from ribosomal RNA sequence data, implies that the major groups of living organisms were all derived from a Last Universal Common Ancestor (LUCA) that formed very early in Earth's history. *Credit:* Chiswick Chap via Wikimedia Commons (CC BY-SA 4.0).

as naturally produced fatty acids could have abiotically formed cell-like structures with semipermeable membrane-like exteriors that give them the ability to concentrate certain molecules in their interiors.[4.2] Processes like this might be able to jump-start the origin of life, and life might commonly form on bodies that provide and maintain suitable conditions. Like biological evolution, the prebiotic processes that led to life were likely complex and involved numerous pathways, many of which led to dead ends or life systems that did not survive. Because all life on Earth uses common processes and compounds, it is widely believed that all life here is directly related to and evolved from a last universal common ancestor, or LUCA for short (figure 4.1).[4.3] LUCA evolved very early in Earth history, but it is reasonable to imagine that there could have been other life-forms that preceded LUCA but did not prevail.

In 1924, Russian biologist Alexander Oparin published his theory of life beginning from a "primordial soup." Though it did include water, this was far from any Campbell's flavor. Oparin had already concluded that the Earth's first atmosphere had been dominated by the simple gases ammonia, methane, and water vapor—and hydrogen and helium, the two most common elements in the Galaxy. Now he proposed that these simple molecules were able to combine, building up ever more complex species over time. At some unspecified point, things that you could call "life" appeared. Oparin was unable to test his theory by experiment.

In parallel, English biologist J. B. S. Haldane formulated a formal mathematical treatment of evolution, starting from scratch.[1] In keeping with his logic and mathematics, Haldane thought of evolution as a series of logical steps, one following the other. He was not necessarily an atheist but a believer that logic and science could explain everything. With no need for outside influence such as a deity, his popular books on this subject enraged the religious leaders of the time. This outlook on life and the Universe is labeled "scientism." Apparently, Haldane was also among the first to foresee a carbon-free energy generation future. In a 1923 talk, he pointed out that coal supplies would eventually be depleted and that England eventually would have to go with some coal-free energy sources, such as wind power.

The first steps involved the prebiotic evolution of complex carbon-based molecules and development of reaction pathways.[4.4] This would then have led to the first living things, our original ancestors. Fossil evidence has been found, including structures over 3.5 billion years old called stromatolites, of what appear to have been extensive microbial mats made by photosynthetic anoxygenic bacteria. Comparing with Earth's age of a little over 4.5 billion years, organisms have existed for at least three-quarters of Earth's history (figure 4.2)!

When life began on Earth, it did so in an atmosphere containing no free oxygen. Without free oxygen, there would be no

FIG 4.2. Five-hundred-million-year-old stromatolites exposed at the Petrified Sea Gardens National Historic Landmark, Saratoga Springs, New York. These are not fossils but structures made by mat-forming bacteria. *Credit:* Photograph by J. Bret Bennington.

ozone to shield Earth's surface from deadly irradiation by ultraviolet light from the Sun. Our earliest atmosphere may have contained modest amounts of free hydrogen and even ammonia (NH_3), but these components could not have been retained for long because they were either decomposed or lost to space. Our early atmosphere surely contained nitrogen, water vapor, and probably significant amounts of carbon dioxide. CO_2 is only a trace gas now, but it was much more abundant in the past. A carbon dioxide–dominated atmosphere is probably a common composition for many Earth-like planets that do not contain either abundant organisms that can modify the composition of the atmosphere or something analogous to Earth's land/ocean/air processes that remove CO_2 from the air and sequester it underground in massive carbonate deposits.

This carbon-bearing gas is the dominant component of the atmospheres of our neighboring planets, Mars and Venus. Over billions of years of evolution, our atmospheric CO_2 has generally declined, although in recent times the trend has reversed due to CO_2 released from burning fossil fuels. The CO_2 abundance is currently 0.04 percent and rising, but this is a small amount compared to what it was in the distant past. Our planet has vast quantities of carbon that was previously in the atmosphere but is now bound up in limestone. Ultimately, as the Sun becomes too bright, this greenhouse-gas time bomb will reenter the atmosphere.

The early atmosphere also contained methane (CH_4), the main component of natural gas. Methane appears to have persisted as a minor component until an amazing transformational event happened 2.4 billion years ago. The watershed event was when methane disappeared, and oxygen began to accumulate in the atmosphere. Carbon in the atmosphere changed from being in a reduced state bonded to hydrogen to an oxidized state bonded to oxygen. The transition between a methane-containing atmosphere and one with free oxygen was one of the most biologically significant events in Earth's history, and it is called the Great Oxidation Event, or GOE.[4.5] It has also been referred to as the Oxygen Catastrophe, the Oxygen Revolution, and the Oxygen Crisis. Free oxygen and methane are chemically incompatible in the atmosphere, because methane, in the presence of oxygen gas, becomes oxidized to form carbon dioxide. Photosynthetic microorganisms had probably been releasing oxygen for at least a billion years, but this gas could not accumulate because, as quickly as it was produced, free oxygen was removed by a series of processes, including reaction with iron that was dissolved in the ocean during that era in Earth history.

When oxygen finally rose in the atmosphere, iron in the ocean became fully oxidized and could no longer stay in solution and act as an effective means of extracting oxygen. Many of the Earth's great iron ore deposits were made before oxygen rose in the atmosphere. The large red boulder near the entry of the Smithsonian Museum of Natural History is a classic example of a banded iron formation, or BIF, an iron ore that usually formed before the atmosphere contained abundant free oxygen. BIFs formed in locales where ferrous iron (Fe^{++}) was dissolved in anoxic sea water and then precipitated to ultimately form layered mixes of silica and iron oxides that contain both Fe^{++} and fully oxidized Fe^{+++}. Oxygen began to accumulate, and methane disappeared only when the supply rate of oxygen exceeded the rate that it was removed from the atmosphere.

Small amounts of methane in the atmosphere may have played a major role in keeping the early Earth warm enough that its oceans never froze over for extensive periods of time. Methane is a potent greenhouse gas and even minor amounts can provide significant warming effects. In addition to methane, the early atmosphere likely contained substantial amounts of CO_2. The young Sun was fainter than it is now, and extra warming from CH_4 and CO_2 may have been critically important to the early history of the Earth. The fact that oceans did not freeze over for extended periods of time has been viewed as somewhat of a mystery that has been called the "faint sun paradox." As a giant nuclear fusion reactor that generates energy by turning hydrogen into helium, the Sun becomes about 10 percent brighter every billion years as its nuclear core temperature rises with age. When the Earth was young, the Sun was a least 30 percent fainter than it is now. The solution to this paradox may, in part, be due to increased warming, sometimes called the greenhouse effect. Small amounts of methane may have kept the Earth unfrozen and habitable for microorganisms, but the loss of methane after the Great Oxidation Event may have triggered instabilities in the atmosphere that led to several short-term freeze-overs, known as Snowball Earth episodes, when sea ice extended down to equatorial latitudes. The sudden loss of greenhouse warming from methane may have triggered the earliest Snowball Earth episodes.[4.6] Carbon in the atmosphere, both as methane and carbon dioxide, probably helped and hampered early life on our planet.

The Great Oxidation Event was a critical event in Earth history, opening the door for the future evolution of organisms capable of aerobic metabolism, an energetic process that eventually led to multicellular organisms, animals, and people. The rise of free oxygen at the GOE also had a downside, and, as noted, the event has also been called the Oxygen Catastrophe. Organisms that could not evolve to survive in the new and dangerous oxygenated world probably became extinct, and the GOE might have caused one of the greatest mass extinctions in Earth history. Microbes are hard to extinguish, but the rise of oxygen, a reactive toxic gas that tends to oxidize or destroy many organic compounds, must have been a formidable challenge for organisms that could not hide from it or evolve mechanisms to live with it.

The evolution of organisms on our planet, the connections to complex planetary processes such as the Great Oxidation Event, and the removal of atmospheric carbon dioxide to form buried carbonates provide valuable insight into the evolution of life on similar planets elsewhere in the Universe. The sudden switch from a methane-bearing atmosphere to one with free oxygen might be a common evolutionary event for other planets that are sufficiently Earth-like and harbor organisms that extensively modify their planet's atmosphere beyond what a lifeless planet would have.

A fascinating question regarding the origin of life is whether the first living (i.e., replicating) organic molecules were nucleic acids or proteins, since we need the nucleic acid to store the self-replication information, and we need the proteins to carry out the chemical reactions. The current thinking favors that the first "living" molecules were probably RNA—but we'll look at DNA first.

Both proteins and nucleic acids are made up entirely of hydrogen, nitrogen, oxygen, phosphorus, sulfur, and carbon. All life on Earth depends on DNA. This large and very complex molecule carries genetic information from one generation to the next as plants and animals reproduce. Human DNA molecules are a few meters long and have some ability to self-repair damage done by radiation or chemical effects. Incredibly, if all the DNA molecules in a human body were stretched out to form a string, it would be over fifty times longer than the distance between Earth and Jupiter. Every creature (except identical twins) has its own unique DNA reflecting past generations.

DNA molecules are long pairs of polymers that are composed of only four types of subunits, called nucleotides: adenine, guanine, cytosine, and thymine. The particular order of these four nucleotides specifies, in three-unit combinations, the order of the twenty amino acids that make up our protein molecules. The DNA nucleotides interact in pairs, so that each strand of DNA has a partner strand, forming the famous "double helix." The DNA is reproduced by being unzipped, so that each side of the pair is used as a template for each half to get a new partner, thus reproducing the entire strand. All of these actions are carried out by proteins. The assembly of amino acids to form proteins and their coding by DNA were mentioned in chapter 2.

RNA is similar to DNA except that one of the subunits is different; RNA is involved in the translation of the information from the DNA into proteins. The amino acids are the building blocks of proteins, and the smallest amino acid is glycine, whose chemical makeup is NH_2CH_2COOH. Glycine was almost certainly present when the Earth formed, because it is present in ancient meteorites and even in samples returned from a comet. Comets and meteorites have delivered glycine, as well as many other organic materials, to our planet over its entire history. It is inevitable that this free delivery system has also delivered glycine to all bodies that orbit the Sun.

If glycine is likely to exist beyond the solar system, it may be detectable in interstellar space. And, like almost any other molecule, glycine can emit radio waves at certain frequencies. These radio waves are produced by rotational transitions.[2] In 2003, a group of astronomers reported the detection of glycine in an interstellar cloud, causing much excitement. However, doubt was cast on this finding in 2004 when a more careful search was made, with no detection. In 2023, it was reported that tryptophan, one of the twenty amino acids essential for the formation of key proteins for life on Earth, had been found in the IC348 star-forming system. The identification was based on the observation of twenty emission lines with the Spitzer infrared space telescope. The search will go on, because finding amino acids in interstellar space shows that this important molecule has a presence in the Universe outside our solar system.

The history of our planet and the slow and difficult evolutionary processes that led to complex life are the basis for the book *Rare Earth: Why Complex Life Is Uncommon in the Universe* by paleontologist Peter Ward and astronomer Don Brownlee.[4.7]

The Rare Earth Hypothesis suggests that life elsewhere may frequently evolve in a somewhat similar manner to what has happened in the solar system. The basic idea is that microbial life may be common in the Universe but that life analogous to Earth's animals and plants is likely to be rare. This proposition is based on the history of life on Earth and also the range of environments that exist on bodies in the solar system. In the solar system, several bodies provide subsurface environments where certain robust microbial organisms found on Earth could plausibly survive, at least for modest periods of time. In great contrast, there is no place in the solar system, other than Earth, that presently provides surface environments that would allow any of Earth's animals to survive. Even the surface of Mars, the most Earth-like planet in the solar system, is a highly hostile environment where no known type of plant or animal could survive. Microbial life began early in Earth history, and it has flourished here for billions of years, in spite of major changes in our planet and even with the Sun.

The Earth experience suggests that the origin and survival of microbial life might be easy as long as a planet provides suitable environmental conditions that do not prohibit its origin, survival, and propagation. On Earth, microbial life developed nearly as soon as the planet's environmental conditions settled down to the point where robust microorganisms could survive. In contrast to microscopic organisms that can survive in extremes of temperature, pressure, acidity, salinity, water content, and other factors, animal life requires quite refined and limited environmental conditions, including the presence of oxygen in the atmosphere. Microorganisms that can live in extreme conditions are called extremophiles, and they have been found in an impressive range of what we consider to be quite hostile environments. The extremophiles give people hope about finding extraterrestrial life—not necessarily monsters as envisioned in science fiction novels, but microscopic things. Some extremophiles exist far below the Earth's surface in environments that must be similar to the subterranean locales of other planets and moons. Extremophiles are impressive, but most have evolved methods to live in a limited range of extreme environments. There are no "super extremophiles" that survive in a wide variety of different extreme environments. For example, you can't expect a microbe that thrives in a boiling hot spring to survive in ice, acid, concentrated brine, on top of a mountain, inside basalt, or as gut flora in the digestive tract of a mammal.

Microbial life is extremely abundant on Earth, with billions of organisms contained in each gram of soil. It is also somewhat extinction proof on a global scale, in contrast to most animal species that commonly become extinct after only a few million years. Microbial life is tough and adaptive, and it formed very early in Earth history. In stark contrast, it took about 4 billion years of planetary and biological evolution for multicellular animals to become prominent enough to leave abundant, clear fossil records. The dramatic appearance of animal species in the fossil record is called the Cambrian Explosion, and it occurred about 530 million years ago.

The solar system experience suggests that while typical stars may have a few planets or moons that could harbor microbial life, the more restrictive conditions and stability needed for animal life are likely to be as rare elsewhere as they are here. Because it took billions of years for animals to evolve on our very environmentally hospitable planet, it is easy to imagine that the development of complex animal-like creatures would be slow or, in many cases, never happen on other planets. Even environmentally ideal planets might never make the jump from microbes to multicellular creatures. The Cambrian Explosion was very late in Earth history, and it occurred after unusual Snowball Earth episodes. Some have conjectured that these odd freezing events may have kick-started the rise of animals on our planet.

The flowering of life after what may have been a fortuitous event is reminiscent of the rise of mammals that occurred after the extinction of the dinosaurs 66 million years ago. Some have suggested that the end of the age of the reptiles, which had lasted almost 200 million years, opened an ecological niche that allowed mammals to flourish. The extinction of the dinosaurs is widely believed to have been caused by the chance impact of a single 10 km asteroid that hit what is now the Yucatán. The impact ejected a globally enshrouding cloud of rock debris and sulfur compounds into the atmosphere.[4.8] We can see that luck, in the form of random events, can play a major a role in the development of animals, and some planets will surely be "luckier" than others.

Moreover, animals and many plants need special conditions and are not nearly as robust as the diverse range of microbial organisms, so it is highly unlikely that life in the Universe will resemble the familiar aliens that inhabit our science fiction books, movies, and television programs. It is most likely that they are highly unphotogenic organisms, individually invisible to the naked eye and only seen collectively as films or what amounts to sludge. Such life-forms have dominated Earth history and are likely to dominate other life-bearing planets.

Microbial life was the first life to form on our planet, and these tiny creatures will be the last to vanish. The Earth's maximum period of habitability is limited by the Sun. Ultimately, it will become bright enough to drive off the oceans and incinerate all life. Eventually, the Sun will expand to the point where it consumes our planet. Earth's overall lifespan will be just over

10 billion years, and microbial life might exist over the majority of this period. Depending on the progress of several processes, our oceans might be lost less than a billion years from now, or it might take longer, but in any case they will be lost. Earth's age of plants and animals may exist for only about 10 percent of our planet's life. From this perspective, animals are rare on Earth over its full lifetime. If aliens came to the solar system ten times over the Sun's total projected lifetime and visited each planet, they would find that only 1 percent of the planets visited would have creatures that were larger than microscopic size. In this timely view, only 1 percent of the Sun's planets would be animal habitable when visited.

Planets are common around stars, but those that can support advanced life are surely rare. How rare is an open question. Is it 10 percent of planets, 1 percent, or perhaps much less? If habitable planets are too rare, then we might not have any that are close enough to us for detailed study. If they are too far away, we may never be able to learn much about them or to see if they contain intelligent life that sends us radio signals. The abundance of habitable, let alone inhabited, planets is one of the great questions to be answered by the emerging science of astrobiology. One of the major questions is how Earth-like a planet really has to be to contain life comparable to what we have here. It is also totally unknown how we fit into the spectrum of inhabited planets. Are we one of the best, just middle-of-the-road, or one of the worst? Does Earth have an optimum amount of carbon or is it advantageous to have either more or less? Also, quite unknown is the long-term survivability of microbial life. While Earth's plants or animals might not survive long, it is plausible that microbes could survive late into Earth history.

On Earth there have been many factors, in addition to having water and carbon, that have enhanced long-term habitability for advanced life. An obvious example is Earth's Moon, an unusually large satellite in comparison to Earth's own size. The Moon's gravity helps to stabilize the Earth's spin axis, thereby keeping the seasonal temperature swings relatively unchanged over long periods of time. Earth's stability contrasts with Mars, whose spin axis flops dramatically and changes the way that solar heating is distributed on the planet. Mars has two moons, but they are too small to stabilize the planet's spin axis. If a species developed at mid-latitude on Mars and had adapted to survive seasons when its tilt was small, it would be greatly challenged when the planet was tilted by a large angle that caused more severe seasonal changes. When a planet's spin is highly tilted, more solar energy per orbit can land on the polar regions than on the equator! An unstable planet will have more extreme long-term climate variations that can harass life and drive into extinction species that cannot adapt to the changes. Our large Moon also raises tides so that the ebb and flow of water in tidal pools alternately moistens and exposes organic material to the atmosphere. While the presence of a large, stabilizing moon may not be critical to the evolution of life, it surely has been a positive factor that has enhanced our habitability and reduced the extinction rate of plant and animal species. Our relatively large Moon, nearly a third of Earth's diameter, is widely believed to have formed because of a collision between Earth and a body the size of Mars.

There are many other factors that have played important roles in promoting the evolution of life on Earth, and the successful evolution of animals and plants. Examples of these include Earth's distance from the Sun; land; water; a life-friendly atmosphere; the presence of plate tectonics; and the relatively low rate of giant impacts from space. The importance of distance is well demonstrated by the nature of our neighbors.

Venus is too close to the Sun and much too hot, while Mars is too far from the Sun, it has no long-term ocean, and its atmosphere is far too cold for the survival of Earth-like animals and plants. Earth's distance from the Sun falls within an annular range of distances from the Sun where surface temperature allows the long-term existence of surface water, something that cannot happen on Venus and hasn't happened on Mars for billions of years, if ever. This range is called the habitable zone (HZ), and many stars are likely to have at least one planet located within this zone.[4.9]

A quite mysterious factor that has been favorable to life on our planet is the process of plate tectonics. It carries sea-floor carbonates deep enough into the Earth that they decompose and release carbon dioxide gas that eventually works itself back into the atmosphere, much of it through volcanoes. This CO_2 source drives the carbonate-silicate cycle, described in detail in chapter 8, that has operated on Earth for billions of years but does not appear to have occurred on other planets. The process could be common on Earth-like planets with oceans, or it might be rare; we have no evidence. It is also plausible that a planet without plate tectonics still might be capable of running a geological carbon recycling process purely by volcanic processes not related to plate movement. The carbonate-silicate cycle occurs on geological time scales, but it acts as a thermostat that leads to warming when Earth enters a long-term cooling trend and cooling when it enters a warming trend. This negative feedback has helped keep Earth reasonably stable. Stability is of great value to complex life because it reduces the extinction rate of species. While extinction events can promote diversity of life, they can also reduce diversity if they occur too frequently. For example, imagine that "dinosaur-killing" asteroid impacts occurred every ten thousand years instead of the expected rate of once every 100 million years. If this were so, humans would have become extinct ever since the glacial ice that covered Seattle, Chicago, and New York last melted.

Earth is clearly a charmed planet in the sense that many factors played roles in its ability to develop and support life. And in this regard, our planet is drastically different from all of the other bodies in the solar system because they cannot support surface life as we know it. After primitive life formed on Earth, it progressed through a long microbial age and eventually evolved animals and plants and abundant oxygen in the atmosphere. Great numbers of species appeared and evolved over time, and most became extinct. Major extinction events followed by the appearance of new species in the fossil record have provided our basis for dividing Earth history into geological periods. As mentioned previously, one of the most famous period boundaries was 66 million years ago, when the age of reptiles ended and mammals became major players among Earth's inventory of creatures. Our primitive human ancestors appeared only in the last 3 or 4 million years. Once the development of intelligence had provided the ability to modify the environment, the entire world became humans' ecological home, and some think that our physical evolution essentially stopped, as we did not have to evolve to adapt to different, and/or hostile, environments. It remains to be seen whether future natural ecological pressures will continue to promote evolution of the already highly evolved human species.

Trying to understand the origin and evolution of life is one of the grandest and most difficult scientific endeavors. In the 1950s, scientists began to conduct experiments in which they attempted to reproduce the conditions of early Earth. The starting point of these experiments was to place water in a container filled with gas that might have been contained in the early atmosphere and add energy to break chemical bonds. The first of these experiments was performed in 1953 by American scientists Harold Urey and Stanley Miller, who concocted an atmosphere of methane, ammonia, and hydrogen in a flask that also contained an "ocean" (water).[4.10] Electrodes above the water produced sparks simulating lightning discharges in the primitive Earth atmosphere. The sparks provided energy to break chemical bonds. After a week, the Miller-Urey mixture turned yellow and eventually a dark brown. When opened, the mix of newly produced chemicals in such an experiment had somewhat of a food smell, in remarkable contrast with the nasty ammonia smell of the original ingredients. Urey and Miller analyzed the composition of this brown "gunk," and found large quantities of amino acids! With a very simple apparatus, they had taken simple molecules and produced some of the fundamental building blocks for life. Other experimenters later showed that exposure to ultraviolet light produced very similar results. And nearly twenty years later, extraterrestrial nonbiological (also known as nonproteinogenic") amino acids were found in a carbon-rich meteorite that landed in Australia. These experiments and phenomena indicate that at least some of the fundamentally important molecules needed for life could have been made either on the early Earth or elsewhere and delivered from space, before any life existed on this planet.

The amino acids produced in the Miller-Urey synthesis differed from ones involved in life in that they were racemic—that is, they contained roughly equal amounts of left- and right-handed amino acids. In contrast, the twenty amino acids produced and involved in life are almost entirely left-handed. The property of molecules to be left- or right-handed is called chirality. An object or molecule is chiral if it can be distinguished from its mirror image, similar to the way your left foot is readily distinguishable from your right foot and will not fit in the wrong shoe. Due to their formation and utilization processes, life is based on and requires chiral molecules. (Molecular chirality was discovered in 1848 by a young Louis Pasteur.)

Now things get a little trickier. Somehow, molecules of the right stuff have to be made by process that could make amino acids, protein, nucleosides, RNA, and other classes of molecules used in life. This is why the initial conundrum of whether the first life-forms were made of nucleosides or amino acids (i.e., RNA or protein) is now favoring RNA, that is, nucleic acids, because they carry the information for their self-replication, and because it is now known that RNA molecules can have catalytic activity and can cause the necessary chemical reactions.

All the intricate chemical processes that make complex molecules have to be sequenced in the proper order and orientation. Some suggestions have been made about how this might have happened—for example, it has been proposed that the correct alignment could have initially occurred along crystalline structures on minerals such as quartz or certain clays—but the actual mechanism remains a mystery. Considerable effort is going into improving our understanding of how life formed from prebiotic compounds. One school of thought is that critical events may have occurred near rare hot-water vents on the ocean floor or hot springs on land. These special environments may have abiotically produced numerous types of macromolecules that then naturally became encapsulated in cell-like membranes. The concentration of materials inside such membranes led to the evolution of primitive life-forms and metabolic pathways.

Here Fred Hoyle weighs in yet again. To disparage the idea of biological beginnings on Earth, and to support his steady-state theory, he proclaimed that there hasn't been enough time since Earth was formed to build life from atoms, no matter how abundant. After all, RNA has hundreds of atoms; if they collide by random movement, even in an environment where atomic motion is happening everywhere, it would take a lot of time, longer than the Earth's lifetime, to build complex molecules.

Hoyle's mistake was that he assumed that only random chemical reactions could occur. This is like the idea that monkeys might be able to randomly type any masterpiece, given enough time. But in the case of Earth's evolution, the Universe had an idea of which keys to tap, and all the time needed to create the ultimate masterpiece. When the first primitive molecules were present, they were able to combine through their ability to selectively bond, creating even bigger and more complex molecules. The development of life probably involved a sequence of events that led to concentrations of complex organic molecules inside environments that could quicken critical steps in prebiotic evolution of the building blocks and chemical systems needed for life.

For Earth, carbon was the champion life-enabling element. As pointed out earlier, carbon is gifted with many ways to bond with other atoms. The vast majority of molecules detected in interstellar space contain carbon. Yet scientists and science fiction writers have pondered the possibility that life elsewhere might be based on an element other than carbon, and Carl Sagan, urging fellow scientists not to be carbon chauvinists, encouraged consideration of life systems elsewhere that might have evolved without relying on carbon chemistry. As long ago as 1891, the astronomer Julius Scheiner had suggested that silicon could be a basis for life. Silicon also has a rich chemistry with many bonding possibilities. Of the elements that have potential for make alternate life, silicon ranks at the top, but it appears to have some serious shortcomings.

A major argument against silicon life is that it has not been found in the solar system. Despite the fact that the solar system's multitude of planets, moons, asteroids, comets, and even dust are all rich in silicon and have been exposed to almost every imaginable environmental condition, it has not yet developed silicon-based life, even after billions of years. Unlike carbon, which is usually rare in rocky bodies, silicon is a major element in all rock-containing bodies in the solar system. Though keeping in mind Sagan's warnings about chauvinistic views toward carbon, it seems clear that silicon has many apparent major disadvantages compared to carbon, at least in the eyes of us children of carbon-based mothers. Silicon is simply not as dexterous as carbon in making complex molecules, and the list of known silicon-based molecules is much smaller than the nearly unlimited range of carbon molecules. Another difference often mentioned is the comparison of oxidized carbon and oxidized silicon. We can eat steak and potatoes and then exhale carbon dioxide back into the atmosphere, thus providing a means for plants to intake this gas, grow, and output free oxygen. This cycle is possible only because carbon dioxide is a gas. Silicon, however, on a planet like Earth cannot naturally exist in gas unless the temperature is high enough to vaporize rocks. If a creature tried to live by eating rocks, they could only exhale or excrete rocks. The change in the degree of oxidation of biomaterials is also a fundamental process that drives the chemical machinery of biology. In contrast to carbon, silicon exists only in its fully oxidized state in almost all natural earthly environments except in the Earth's quite inhospitable core.

Another shortcoming of silicon as a basis for life is the lack of an adequate solvent. Terrestrial life critically depends on water and its remarkable properties, such as its ability to dissolve certain compounds, diffuse through membranes, and move materials around. Water at moderate temperatures turned out to be just right for carbon-based life, providing transportation—a means by which to get materials into cells and take waste out—and a medium so that life's molecules can replicate themselves and control life's most basic functions. But in a silicate rock, there is no short-term movement unless it is melted. In contrast with organic compounds, silicon compounds are not soluble in solvents that are likely to be present near the surface of an Earth-like planet. Rocks do dissolve by weathering or when in contact with hot, pressurized water, but this is on a vastly slower time scale than the quick pace of life as we know it.

Some people imagine that there might be drastically different planets that might be hospitable to silicon-based life, but until it is found—perhaps on a planet that contains hydrofluoric acid or pressurized water hot enough to dissolve rocks—silicon-based life will remain an unlikely possibility in the minds of most of the astrobiology community, though our views of the environmental needs of life might change as we more fully explore some of the solar system's more exotic places. A good example of a place that might harbor life but is quite different from Earth is Titan, the mysterious moon of Saturn that has lakes of liquid hydrocarbon on its surface.

So, the only life that we know of is carbon based. Averaged over geologic time, invisible microbial life has been the major form of life on our planet. It is possible that microbes, as their environmental requirements are simpler than those of multicellular life, are the most common form of life in the Universe as well. Microbial life preceded plants and animals here, and it is surely quicker to form and easier to evolve. Microbial life has been found in places where multicellular animals cannot survive. Earlier we mentioned critters called extremophiles. Probably the best-known extremophiles are those found in hot springs on the surface and in deep-sea vents. The microscopic extremophiles in hot sea-floor vents can support an amazing ecosystem that contains crabs and tube worms living in the cooler water surrounding the superhot water streaming out of the vents. In the deep ocean, these creatures do not need light or photosynthesis, though they do require ingredients such as oxygen that come from the atmosphere.

The existence of extremophiles gives people optimism about finding extraterrestrial life. Some extremophiles exist far below the Earth's surface in environments that must be similar to subterranean locales on other planets. For example, the subsurface environments on Mars and many moons are very similar to those on Earth. Of course, we don't know about everything that carbon-based life needs to begin and carry on, but we know that it requires a liquid medium where molecules quickly move around to meet other molecules; food; and a source of energy like a star nearby. Water fits the bill as a liquid medium. Lighting and heat from the interior could be the energy source, and carbon is part of the food, along with hydrogen, nitrogen, and oxygen. The Sun provides the energy to power life here, but it is possible that life elsewhere, and even some subsurface regions on Earth, could be powered by chemical energy sources, such as the oxidation of iron in basalt.

The Earth is inhabited by a broad range of life because it has an atmosphere that has maintained habitable conditions for billions of years. We have a favorable atmosphere, and our planet is in the Sun's habitable zone (HZ), the range of distances from a star where an Earth-like planet can retain liquid on its surface (figure 4.3). We have talked about the importance that distance from the Sun plays in allowing the necessary chemical and physical processes that gave rise to life on Earth. Outside the HZ, surface water can freeze, and interior to the HZ, water can be lost to space. The location and extent of the habitable zone

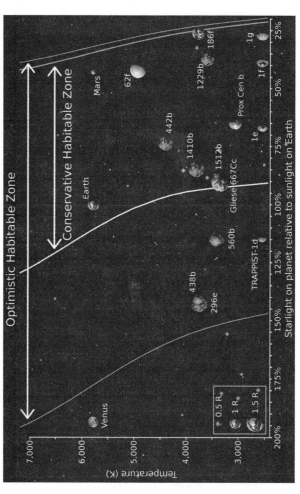

FIG 4-3. The Habitable Zone, locations where planets might have surface oceans. Conservative and optimistic estimates for cool, lower-mass and hotter, higher-mass stars. Exoplanets that appear to be in the habitable zones of cooler stars are labeled in the region below Venus, Earth, and Mars. In this graph, the intensity of starlight is plotted instead of the distance from the star. *Credit*: Chester Harman; PHL at UPR Arecibo, NASA/JPL.

depends on the type of star and also on the planet's orbit, age, and atmospheric composition (and greenhouse heating); the stability of the host star; and whether the star is part of a binary or multiple star system. A very hot star has a broad habitable zone far from the star, while a cooler star has a narrower zone closer in. If planet distances are distributed logarithmically, like in the solar system, then a star might have several planets in its habitable zone. Some systems are tightly packed. The TRAPPIST 1 system has seven planets similar to Earth orbiting very close to a low-mass star, and three appear to be in the habitable zone. All seven planets are packed into an annular ring only 10 percent of the width between the orbits of Earth and Mars, so if any of the planets have life, they might also have very close neighbors with life on them.

You might think those stars with the largest habitable zones would be the winners in the search for life, but there is another important consideration: the life expectancy of the star. The stars with the largest habitable zones are hot stars, which burn out relatively quickly. The most massive and the hottest stars have very short lifetimes before they explode as supernovas or disappear in black holes. Instead of billions of years, the hottest stars go supernova in less than a million years. Any primitive life-forms, if they were there, would not get much time to evolve into more complex beings. Given that it took about a billion years for the first primitive life-forms to develop on Earth, scientists generally rule out stars much hotter than the Sun as likely hosts for complex life.

Another factor is that stars' properties change as they age. Even if a planet is outside or inside the habitable zone now, it may not have always been that way. The Sun is increasing its luminosity as it converts hydrogen into helium in its interior. For most of the Sun's life, it becomes about 10 percent brighter every billion years, as noted previously. In the 5 billion years since the planets formed, Venus may have been in the Sun's habitable zone when the Sun was less luminous, and Mars has probably been in the habitable zone since its formation. The words "may" and "probably" are used here because the inner and outer boundaries of the zone cannot be predicted or determined precisely, due to the possible range of complex processes that plausibly could occur on Earth-like planets over their lifetimes. If we could move Earth to different distances from the Sun, we could see what would happen over a billion-year time scale. When the Sun finally becomes a red giant star, Pluto might have liquid water on it, or least close to its surface, while Earth and Mars get roasted. It is the fate of all planets that form in the habitable zone that eventually their stars become too bright.

What about young stars? Newly formed stars undergo a phase when they are brighter than they will be after they settle down and become normal stars. This "adolescence" is mostly an issue for stars less massive than the Sun, and young planets around such stars might get prebaked before they can bask in nominal HZ conditions that could last for billions of years.

Our Sun seems to be just right: its habitable zone is fairly broad; it has a long lifetime; and it is stable, not emitting many flares or pulsating like some stars do. We can consider the Sun to be like the three bears and the bowls of porridge: one is too hot, one is too cold, but the Sun is just right. (Though if Carl Sagan were reading this, he probably would immediately point out that this is not necessarily correct, because we Earthlings usually view such matters from a biased, Sun-centered viewpoint.) In contrast, low-mass stars are cooler than the Sun and have long lifetimes, longer than the Galaxy's age. The lowest-mass stars last nearly forever. Cool stars are quite faint and have very small habitable zones, but can still have multiple potentially habitable planets.

These stars have other problems, though: they emit mostly in the red or infrared part of the spectrum, where the photons have less energy than visible photons, and these stars have a bad habit of emitting frequent energetic flares. When a solar flare occurs, we usually hardly notice it on Earth, but if we lived on a planet around a cool star, the star's numerous energetic events would be life threatening.

In our solar system, let's consider life on our planetary neighbors, Venus and Mars. These planets are not at all like Earth from the standpoint of biological processes. One is much too hot, and the other is too cold, among other factors. No known organism on Earth could survive, let alone be happy, on the surface of either planet.

However, as early as the late 1800s, fiction authors wrote stories about life on Venus. In 1895, Gustavus Pope published a book featuring soldiers fighting dinosaurs. In 1918, Swedish physicist and chemist Svante Arrhenius, whom we met earlier as the promoter of the panspermia hypothesis, thought the clouds covering the planet must be made of water vapor, supporting trees and forming swamps. From there, science fiction authors, including Edgar Rice Burroughs, Olaf Stapledon, Robert Heinlein, Henry Kuttner, and Ray Bradbury, have told tales about life on Venus, either pleasant or dangerous. The stories involved humans traveling to Venus; indigenous swamp creatures (often dangerous to humans); flying bird-like intelligent beings populating the skies; or all of the above. Venus life also featured in a number of movies from the golden age of sci-fi.

Then reality hit. Radar and spacecraft observation in the 1960s showed a very hot, rotating-backward planet that couldn't possibly support life. Furthermore, the clouds weren't made of water but were composed largely of an aerosol of microscopic sulfuric acid. Some other minor gases also play important roles: sulfur dioxide (SO_2), for example, which is also present in Venus's clouds, acts as an absorber of solar ultraviolet radiation in the upper atmosphere. Because SO_2 absorbs ultraviolet radiation, it therefore plays a role very similar to that of ozone in the Earth's atmosphere. In global warming, Venus is subject to a double whammy: both CO_2 and SO_2 trap heat. Both Russia and the US have sent robotic spacecraft that did fly-bys, orbited, and even landed on the hot surface of Venus and persevered for a while before they were finally cooked to death. The landers on the rocky surface found conditions impossible to support life. At the surface, the mean temperature is 464°C/867°F, and the "air" pressure is a hundred times that on Earth! Highly detailed maps made by cloud-penetrating radar show vast expanses covered with lava flows; spectacular, large-impact craters; and tall volcanic mountains. With a barren surface hot enough to melt lead, it is no wonder the landing craft couldn't survive.

As for the vital presence of water, we don't have any information about the early history of Venus, but it is likely that it started out with as much water as Earth. Its dense, nearly pure CO_2 atmosphere shows that it formed with and retained appreciable amounts of carbon; in fact, the Venusian atmosphere contains almost a million times as much carbon as ours. If the carbon in the atmosphere of Venus were converted to solid carbon on the planet's surface, it would be roughly a kilometer thick. As mentioned earlier, the young Sun was fainter than at present, and it is possible that Venus initially had oceans but eventually lost them to space. Being closer to the Sun, the sunlight intensity on Venus is nearly twice that of Earth, and this is probably the fundamental reason that Venus is now a dry planet.

It is commonly imagined that Venus had a greenhouse episode during which the rate of energy captured from the Sun could not be balanced by radiating energy back into space. This positive feedback led to a runaway event in which temperatures may have become hot enough to melt surface rock. The planet's hellish greenhouse history, probably driven by water vapor, caused the oceans to be lost by hydrodynamic escape, the outward flow of atmospheric gases into space. Oceans can also be lost by the photochemical splitting of water molecules at the top of the atmosphere, and some of the freed-up atoms and molecules have velocities high enough to escape.

Current conditions on the surface of Venus are too hot for liquid water, so it is impossible for life as we know it to exist there. None of the organic molecules that are important to life on Earth can survive on or below the surface of Venus. Some scientists, however, have speculated that some different type of chemistry might promote a form of life at these high temperatures. One possibility is that life did arise on Venus during the planet's early history, before its oceans evaporated, and conditions became so hot. Such scenarios are matters of conjecture, but future Venus missions will shed insight into this planet's largely unknown past.

Now let's look at Mars. It has been the subject of speculation that life exists there—not only in science fiction but also in modern science. NASA, China, and Europe are spending billions of dollars to explore Mars and search for life. With astronomical methods from Earth, a good time to observe Mars is opposition, when Earth is between the Sun and Mars, and the planet is closer to the Earth than usual, which happens about every two years. In 1877, many astronomers took advantage of the observing conditions to study Mars. Asaph Hall discovered the two tiny moons of Mars then,[3] and Giovanni Schiaparelli made sketches of Martian surface features, based on visual observations through a telescope.

Others (not Schiaparelli himself) interpreted those linear markings as long canals, carrying water from the icy poles to warmed climes in which crops could be cultivated for the benefit of Martian society. The notion that a Martian civilization not only existed but furthermore was involved in large-scale land reclamation projects took the fancy of many people. Most notable among them was the American banker Percival Lowell, a Boston aristocrat with an abiding interest in astronomy and the financial means to indulge his passion for the science. Lowell was so entranced by the idea of a Martian civilization that he wrote a book, published in 1896, describing his vision of Martian society. In Lowell's imagination, the Martians had an agricultural economy, but their planet was drying up; therefore, the Martians were forced to build canals to transport water from the polar ice caps to the temperate zones where crops were grown. Lowell was not in the least deterred when scientists pointed out that the canals would have to be enormous (over fifty miles wide!) in order to be seen from the Earth. Lowell's thoughts of "canals" were later determined to be erroneous, resulting from the tendency of the human eye and brain to connect points separated by small spaces. Lowell's enthusiasm for life on Mars never waned, although he did find another quest later in life, the search for a ninth planet. Pluto was discovered at Lowell Observatory shortly after Lowell's death.

When transcontinental radio telegraphy became a miraculous reality, at the beginning of the twentieth century, Marconi operators were told to be on the alert for possible faint signals from Mars as well as those from distant ships at sea. The idea of life on Mars inspired many other science fiction tales, even more than Venus. Starting in 1912 with Edgar Rice Burroughs's novel *A Princess of Mars*, life on Mars became a hot topic. Among the many authors who envisioned Martian society were C. S. Lewis, Ray Bradbury, Lester del Rey, Leigh Brackett, Isaac Asimov, Robert Heinlein, and Arthur C. Clarke, to name some of the most famous among them. The phrase "little green men" came from *A Princess of Mars* by Edgar Rice Burroughs, and to this day refers to any alien species. The concept of a Martian society (a hostile one this time) experienced a spectacular renewal in 1938, when Orson Welles staged his famous *War of the Worlds* radio broadcast based on H. G. Wells's earlier (1898) novel of the same title. This simulated news report featured spacecraft landing in New Jersey and a murderous rampage across the countryside by Martians intent on capturing our planet and enslaving its people. The broadcast was done as a series of news flashes, and many listeners, having failed to note the disclaimers, believed the invasion was really happening. Widespread panic swept through New Jersey and New York before the word got around that the "invasion" was created for a radio drama.

General interest in the Martian civilization did not go away until the 1960s, when the first close-up images showed what a desolate place Mars is. More serious speculations about life on Mars were based on observation of annual changes in coloration and on the supposed detection of features due to chlorophyll (a carbon-based molecule) in the spectrum of reflected light from Mars (this later turned out to be a misidentification). But hopes of finding at least plants on Mars faded when the first images from *Mariner 4* revealed nothing but craters and dust. The annual color changes were soon explained by Carl Sagan as being caused by seasonal windstorms, which shift the dust around on the Martian surface, causing changes in the distribution of light and dark areas.

In 1976, the two *Viking* spacecraft sent complex instrument packages to the ground on Mars to look for evidence of life. Although fifty missions have been sent to the red planet, *Viking* is the only successful one with a specific focus on life detection. The landers were highly successful and lasted several years. The costs of two *Viking* landers came to billions of 1976 dollars, a reflection of the high priority on Mars as a place to look for extraterrestrial life.

The landers had four instruments to perform experiments on the soil.[4] One experiment in particular caused much excitement because it found evidence of small organisms living in the soil. Some dirt was scooped up and dribbled into a container of food. Actually, the "food" was a soup that contained ^{14}C, the radioactive form of carbon. After a while, the gas in the container was analyzed and ^{14}C was found in the gas. That meant something in the nutrient was eating and excreting! Unfortunately, the things didn't eat twice, as living organisms would do. In the end, unexpected chemical reactions were thought to be the cause of the excitement.

Certainly, Mars once had better conditions to support life. Ancient river channels, deltas, and putative ancient shorelines are evident, and minerals and formations made by water have been observed by rovers. Although it is clear that there are no plants and animals on its surface, Mars could have microbial communities existing even now, under its surface, in wet, warm environments. Unlike the case for Earth, Martian organisms, if they exist at all, have not produced readily detectable changes to either the Martian atmosphere or surface.

Martian meteorites provide an intriguing insight into the possibility of life on Mars. Over three hundred rocks have been blasted off the surface of Mars and found on Earth. They are identified as Martian meteorites on the basis of their chemical and mineralogical composition, and in some cases the composition of Martian atmospheric gases trapped inside them. Most of these Martian rocks were found in our planet's most productive meteorite hunting grounds, the cold and hot deserts of Antarctica and North Africa. The most famous Martian meteorite is one in Antarctica called Allan Hills 84001 (ALH 84001). With a formation age of over 4 billion years, it is the oldest Martian meteorite, and it contains traces of organic material and carbonates, compounds that we now know are present on Mars but were not abundant enough to be easily detected from orbit or landers before the meteorite was found. ALH 84001 was present on Mars during its early period, when the atmosphere was denser and warm enough that water could flow on its surface. A 1996 paper created a considerable amount of drama and discussion when it made a case that this meteorite contained not only organic materials and carbonates but also features that resembled the fossilized remains of microscopic organisms. If Mars ever had microbial organisms, it is likely that they or their fossilized remains have hitchhiked on the millions of Martian meteorites that have fallen to the surface of our planet. While the concept remains tantalizing, the general consensus is that at present there is no definitively compelling evidence that microbial fossils from Mars have been found in any meteorite.

The search for life on Mars remains a high priority for NASA research. Not only is extraterrestrial life important for philosophical reasons, but also for NASA funding. Imagine what finding life on Mars would do for NASA! Currently the car-size NASA rover *Perseverance* is collecting carefully selected Martian samples that will be returned to Earth for detailed analyses in laboratories that will provide highly sensitive and adaptable ways to look for evidence of present or past life on Mars. Since *Viking*, rovers, landers, and orbiters have provided a much-advanced view of Martian history. Although the surface of Mars is now frozen, it is clear that Mars had surface conditions in the distant past that allowed water to flow.

Mars is an excellent example of how difficult it is to find positive evidence for life on a planet that may or may not have, or have had, microbial life but certainly has never been gifted with life to the extent that Earth has been. We have spent many billions of dollars and half a century of effort in spacecraft exploration of a planet in our astronomical backyard. It is never more than twenty light-minutes away from us, whereas the nearest star is over four light-years and most of the stars in our Milky Way Galaxy are about fifty thousand light-years away. We still can't say if Mars is living or lifeless, but, at the very least, studying the question through what the sixth element can tell us about life on our planet brings us ever closer to an answer.

CHAPTER 5

Carbon in the Milky Way

Now that we have seen how carbon is distributed in the solar system, how about outside it? That's where carbon came from, after all. Interstellar carbon was made inside stars, and then distributed around the Galaxy by explosions and stellar winds, as explained in chapter 1. It's a carbon cycle, but not the one we usually think of: in the atmosphere (CO_2); decaying plants; fires; and carbon going back to the atmosphere. Now we're looking at the cosmic carbon cycle: carbon made in stars; carbon getting out of stars; interstellar carbon; and carbon back in stars and planets.

That there is gas and dust between stars, in interstellar space, was known since the first telescopic images were taken of the Orion constellation, specifically Orion's sword, where there is a fuzzy region, made not of stars but rather of gas—what we call a nebula. Astronomer Edward Emerson Barnard, using the forty-inch telescope in the Yerkes Observatory in Wisconsin, made a series of photos of small regions of the nighttime sky, and published a catalog of fine photographs in 1919. In it, there were dark regions where something was blocking background stars (figure 5.1).

In 1930, astronomer Robert Trumpler showed that distant star clusters appeared fainter than they should. The Galaxy is

FIG 5.1. The dark interstellar cloud Barnard 163 blocking background stars. *Credit:* T. A. Rector/University of Alaska Anchorage, H. Schweiker/WIYN and NOIRLab/NSF/AURA.

permeated with a medium of dust that dims light from stars. In other words, because of small dust particles, the brightness of stars falls off with distance faster than expected from the normal inverse square law. The light-absorbing dust in the space between stars, known as the interstellar medium, or ISM, is clumpy, not smooth. There are bright regions of clouds near stars, whose radiation excites gas atoms until they glow.

Because the particles tend to block out short wavelength light more effectively than the longer wavelengths, red light penetrates the interstellar medium more easily than blue light. Thus, distant stars appear not only dimmer but also redder than they otherwise would. The tendency of the dust to make stars appear dimmer is called interstellar extinction. This trend of increasing extinction with decreasing wavelength continues throughout the spectrum, so that the extinction at ultraviolet wavelengths is very severe, but it is minimal in the infrared and radio portion of the spectrum. Thus, it is virtually impossible to observe the inside of even moderately dense interstellar clouds with an ultraviolet telescope, but we can see deep inside the densest clouds at infrared and radio wavelengths. The Hubble Space Telescope has cameras to make optical, near infrared, and ultraviolet images, as well as obtaining spectra.[1]

Most of the mass (about 99 percent) in interstellar space is in the form of gas atoms and molecules; the other 1 percent is the dust, tiny particles of solid matter. The interstellar medium is analyzed by spectral analysis of its lines and bands. Gas produces sharp spectral lines, while the dust sometimes causes broad features often seen in the infrared. Interstellar carbon is in several forms: from grains, to molecules, to ions.

There are about 10^{21} atoms of hydrogen in a cylinder of a cross-section of one square centimeter that stretches from us to a star 100 light-years away; that's about 10^{-23} grams/cm^3 or a few atoms/cm^3. And those are the numbers for hydrogen, the most abundant element. Carbon is about two thousand times less abundant.

The range of physical conditions in interstellar space is enormous. The densest interstellar clouds have densities of perhaps 10^4 to 10^6 particles/cm^3. Compare this with the Earth's atmospheric density of about 2×10^{19} particles/cm^3. Interstellar space is practically empty—but that's where we come from!

But space is not really empty—there's the interstellar medium. Since we can't see the ISM up close, we have to use telescopes to observe across large distances. Observations are needed across the full spectrum, including X-rays, ultraviolet, visible light, infrared, millimeter wavelengths, and longer-wavelength radio. Carbon emits and absorbs radiation all across the spectrum. Usually special types of telescopes are needed to obtain data at different wavelength ranges. Visible light reaches us, but X-ray, ultraviolet, and some parts of the infrared spectrum are entirely blocked by the atmosphere. Observatories on high mountains alleviate some of these problems, but not all. Observatories in space are the answer, but even the largest space telescopes have much smaller diameters than the new generation of thirty-meter-diameter telescopes that are being built on the ground.

Visible light, ultraviolet (UV) and infrared (IR), mm-wave and radio radiation, and even gamma rays are all electromagnetic radiation and differ only by their wavelengths and frequencies and the energy they can convey. We won't say much more about gamma rays, because they just destroy carbon molecules in interstellar space. Other kinds of radiation interact with carbon in space more gently.

Electromagnetic radiation, wavelength, frequency, and energy are all related by the simple equation

$$c = \lambda \nu$$

where c is the speed of light; ν stands for frequency, and λ is wavelength. Longer wavelength means lower frequency and vice versa. So:

wavelength = the speed of light divided by frequency

and

frequency = the speed of light divided by wavelength

For many purposes, electromagnetic radiation is considered to travel in packets called photons. The energy of a photon with a frequency of v is hv and a wavelength λ is hc/λ, where h is called Planck's constant, a value that never changes. Note the energy is higher when the wavelength is shorter. This explains the damage that X-rays and ultraviolet can cause, because they have enough energy to break chemical bonds.

Light is emitted or absorbed when electrons in an atom move to higher or lower levels. Both occur in the interstellar medium. Every element has its own electron structure; therefore, it can be identified by its spectrum. That applies to interstellar clouds, so we can derive the composition of the cloud by observing its spectrum. We can also tell the temperature in the cloud. In a cold or cool cloud, the electrons are normally in their ground states. When starlight comes through the cloud, some electrons are raised to a higher state by absorbing energy from the light, forming an absorption line in the spectrum. In hot clouds, many more electrons stay in excited states, and when they drop to lower energy states, an emission line is created. Each element has its own pattern of wavelengths, either absorption or emission. Thus, astronomers can identify which element is there—for example, carbon.

ISM clouds far from any hot stars are very cold, with temperatures typically less than 50°K. The less dense, diffuse clouds have densities of one to one thousand particles per cubic centimeter and temperatures of 50° to 150° K. Most of the volume of galactic space is filled by even more tenuous material, with a density as low as 10^{-4} particles per cubic centimeter and a temperature as high as a million degrees Kelvin, caused by supernovae, which occur at a rate of about one every fifty to one hundred years in the Milky Way.

The low density in the ISM permits carbon atoms to get along without encountering other atoms or molecules for some time—as much as several hundred years, depending on the local density. Unlike a planetary atmosphere, where carbon atoms are always in molecules, lone carbon atoms or ions[2] can live for a long time in the interstellar medium. It is only in dense clouds (the ones you often see in pictures) that most of the carbon is contained in molecules, mainly CO (table 5.1).

The Sun is in a void, or the "local bubble," where gas has a temperature of about 7,000°K and a density of 0.3 atoms per cubic centimeter. The ISM, however, has many bumps and wiggles. Its variations of density and temperature produce the "violent interstellar medium," as one of us (TPS) called it in a technical review chapter.[5.1]

Radio, UV, x-ray, and even gamma ray observations reveal high-speed clouds, with speeds up to several hundred kilometers per second, and high temperatures, up to 7,000°K. Between these clouds, the intervening gas is millions of degrees hotter. Carbon is highly ionized; C III and C IV dominate. The sources of the heat are high-speed winds from massive stars and supernova explosions. This stuff fills about 70 percent of the volume of the Galaxy.

If a hot star is embedded inside an interstellar cloud, its radiation often forms a surrounding region of ionized hydrogen, an H II region. These regions glow red due to emission from hydrogen atoms, the most abundant element in the ISM. Color photographs in astronomy, such as those we see on calendars, often show red H II regions. Carbon also has emission lines, but they are weak, compared with the hydrogen ones.

As interstellar clouds contract, on the way to star formation, they cool, and more complex molecules become possible and dust particles grow. The end products are a star or a binary star and planets, as described in chapter 3. Now we are thinking about the processes before that, the dust and gas as the cloud gets its act together to create a star.

TABLE 5.1. Interstellar medium components in the Milky Way

Component	Fractional volume	Temperature (K)	Density (particles/cm^3)	State of hydrogen	State of carbon
Molecular clouds	<1%	10–20	10^2–10^6	Molecular	Molecular (CO)
Cold neutral medium (CNM)	1–5%	50–100	20–50	Neutral atomic	Atomic (C I), Ionized (C II)
Warm neutral medium (WNM)	10–20%	6,000–10,000	0.2–0.5	Neutral atomic	Ionized (C II)
Warm ionized medium (WIM)	20–50%	8,000	0.2–0.5	Ionized	Ionized and doubly ionized (C II and C III)
H II regions	<1%	8,000	10^2–10^4	Ionized	Ionized and doubly ionized (C II and C III)
Coronal gas Hot ionized medium (HIM)	30–70%	10^6–10^7	10^{-4}–10^{-2}	Ionized (metals also highly ionized)	Coronal gas (C IV, C V)

Note: The fractional volume shows the approximate range of volume occurrence in the disk.

When astronomers observe a faraway star, they find gas and dust between themselves and the star. Only the more abundant elements show up, because the less plentiful ones don't absorb enough light to be detected. The ISM elements follow the distribution in Sun-like stars, but the condensable ones end up in dust while the volatile ones remain as gas, as we will describe shortly. In diffuse clouds, normally there is enough UV radiation to free one electron from its nucleus, creating a singly ionized element. Carbon is mostly ionized C II, but a little neutral C I remains. You would think no molecules could live in such a harsh environment, bathed in UV, but a few simple ones remain. The first interstellar molecules discovered, in the 1930s, were CH, CN, and CH$^+$ (the ionized CH molecule gains charge by losing an electron). By far the most abundant molecule is hydrogen (H_2), observed only by UV telescopes. Following H_2 is CO, again having a UV spectrum.

So how do astronomers use radio telescopes to "see" the molecules deep in the cloud, where visible and UV wavelengths don't usually escape to be seen? A dark cloud, by definition, has almost no starlight passing through to form an absorption spectrum. That is all right, though, because most elements are in molecules, which have rich radio spectra, and radio wavelength radiation has no problem seeing into dark clouds. The first astronomical radio telescope was built by Grote Reber, a radio amateur, in 1937. With the great radio and electronic advances made during World War II, radio astronomy changed dramatically when the war was over. The first radio observations of interstellar molecules came in 1963. That was a simple molecule: CH (methylidyne) (figure 5.2).

More complex species have been identified in dense, dark interstellar clouds, and more are being found frequently. The more complex molecules, the ones having four to eleven

FIG 5.2. The structure of a CH (methylidyne) molecule; the first interstellar molecule discovered by a radio telescope.

atoms, are observed in dense clouds, mostly in star-forming regions. The molecular zoo is getting bigger with each new identified species, and currently there are hundreds.[3]

Now we get to the other 1 percent. Interstellar space has tiny dust grains mixed in with the gas, as deduced from the way light goes through clouds. This solid matter forms in the outflows from giant stars in their dying years. Hot star outflows are called stellar winds, with speeds as high as several thousand kilometers/second.[5.2] In the much more common cool giants, the winds are slower, only tens of kilometers/second. Perhaps we should call them stellar breezes, rather than winds. (But note that even that speed is much higher than supersonic airplanes!)

Another important source of interstellar dust is debris from explosions of massive stars: supernovae (SN). As the gas from the SN cools, atoms and ions condense, forming tiny grains. The dust particles are small, ranging in size from a fraction of a micron (one millionth of a meter) to something big enough to see with a naked eye. The size of the grains dictates the wavelengths of light that can pass through a cloud. The very small grains affect UV and blue radiation the most, and the largest grains don't do much except emit radiation in the infrared and radio wavelengths. Radio waves pass through the dust in dense clouds and enable astronomers to "see" radio spectra of molecules. There are two

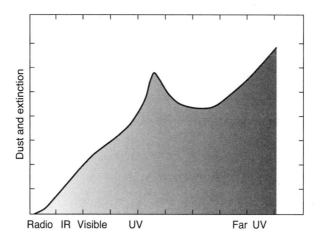

FIG 5.3. The interstellar extinction curve showing the wavelengths of light that have trouble getting through long path lengths from faraway stars. The vertical axis is a logarithmic scale indicating how much flux from a source has been absorbed. The horizontal scale shows the wavelength from radio to the far ultraviolet. The extinction per distance on the vertical scale ranges by about 20 million from radio to ultraviolet. The bump in the center is the 217.5 nm ultraviolet feature caused by interstellar organic molecules, either on dust grains or free molecules.

distinct kinds of grains. The larger are mostly silicates, oxygen/silicon/magnesium/iron minerals such as pyroxene and olivine or noncrystalline materials. The smaller grains are made of carbon-rich materials such as graphite, amorphous carbon, silicon carbide, or PAHs.[4]

Observations of distant stars affected by dust are made by accounting for the particles' spectra, evaluated by an "extinction curve" (figure 5.3), which can then be compared to the spectrum of a similar star that is not affected by dust attenuation. The measured extinction curve slopes generally upward, with a significant bump in the UV region and a high rise going to the far UV region. The IR to the visible regions of the curve is formed by the larger silicate dust, and the bump and the rise by the smaller grains, made of carbon. That the longer wavelengths easily get through the ISM, while the shorter wavelengths are stopped, explains the reddening of faraway stars, as noted earlier. The prominent ultraviolet bump rising above the general curve is attributed to graphite, or a close relative, because carbon atoms in a lattice always absorb UV at a specific wavelength. Graphite is not the only candidate; other forms of carbon, like polycyclic aromatic hydrocarbons, could do this.

The elements, if they can, cling to or become part of dust grains, a process that causes them to become depleted from the gas that is largely hydrogen and helium. For example, iron's depletion is about 99 percent, meaning only 1 percent stays in the gaseous interstellar medium. For calcium, the depletion is 99.99 percent—there is almost no free calcium in the ISM gas. Carbon's depletion is about one-half to one-third, meaning there is enough carbon left in the gas to form complex gas molecules, which it does.

There are many interstellar molecules that remain unidentified. A young Berkeley graduate student, Mary Lea Heger, in an obscure list of interstellar spectral lines, found two unidentified features that seemed to be stationary, not moving with a star. They were rather indistinct spectral lines, difficult to measure precisely, and she made no particular mention of them apart from including them in her list. Once she had completed her doctorate in 1924, she never returned to research. But her passing reference to a couple of "stationary" lines in stellar spectra introduced one of the premier astronomical mysteries of the twentieth century—and now the twenty-first century as well.

For more than a decade, no attention was paid to the two spectral lines lurking in obscurity in a table of numbers in a short paper published in a relatively minor journal. But then another Lick/Berkeley graduate student, Paul Merrill, took note of Heger's list and decided to do some follow-up observations. Merrill, too, had been working with stellar spectra, but he paid more attention to the stationary lines. He found more of the mysterious features, and in the mid-1930s he published a series of papers specifically about them. He called them unidentified diffuse interstellar bands (DIBs).

By now, astronomers have found more than six hundred of the mystery bands, mostly in the yellow-red portion of the spectrum. We don't know what forms them—but we know a lot of things that don't.[5.3] Many hypotheses have been put forward, and many have been shot down. Dust particles in space, frozen oxygen molecules, hydrogen molecules in special energy states, and negative atomic ions—all have been ruled out by astronomical or physical evidence. Chemists have joined the chase, and biologists, or at least those following the new subdiscipline called astrobiology, have become involved as well.

The entire problem of the mysterious bands has taken on an importance that goes beyond the specialized challenge of explaining a few unidentified spectral features. Now the problem of the DIBs has achieved cosmic status, with possible implications for the origin of life itself. Why? Because the hypotheses that have survived every challenge involve carbon. The diffuse bands seem to be telling us that complex carbon-bearing molecules are widely dispersed throughout space. Before the 1990s, the DIBs had little respect—they were just a curiosity before the carbon hypothesis came to the fore. Now they are recognized as very important, because up to 15 percent of the carbon in the ISM is jailed in these molecules, whatever they are.

No known molecules, either small or large, have spectra anything like the DIBs. The molecules are probably big, because simple ones always have simple spectra, unlike the six hundred–plus DIBs. Diffuse interstellar bands are now known to exist on the edges of interstellar clouds, not the denser interiors. The ultraviolet radiation from hot stars permeates the ISM outside of dense clouds, so the bands' molecules are probably ionized, with an electron missing, making the molecule more reactive.

A candidate group for the mystery bands is chain molecules, with carbon atoms in a line. They have been detected in dense, dark interstellar clouds by radio telescopes. In diffuse clouds, where the DIBs live, only C_5 has been found. There was some excitement when C_7^+ (ionized C_7) was proposed as a source of the bands. That was based on a match between lab and interstellar DIBs. But when other observers went to work to test this, they found that the match wasn't good enough.

Another group of molecules, the polycyclic aromatic hydrocarbons, or PAHs, described in chapter 2, are also very good candidates (figure 5.4). One problem is that they come in a lot of shapes and sizes. There is no limit to how many six-carbon rings could be attached together, and a PAH molecule can stretch and bend, forming different spectra in all forms. If every PAH produces its own DIB, the identification may be impossible, or at least it would take a very long time.

One pair of DIBs, in the far-red spectrum, have been linked to a source: a buckyball, in its ionized form, without one electron. Its symbol is C_{60}^+. Buckyballs, or buckminsterfullerenes, also described in chapter 2, have simple spectra, and they don't match hundreds of DIBs.

So the mystery of the DIBs remains a premier problem in astronomy. The most complex interstellar molecules, containing more than four atoms, live in the densest clouds, where atoms and small molecules can collide and stick together. As of this writing, the largest identified molecule is the fullerene C_{70}, men-

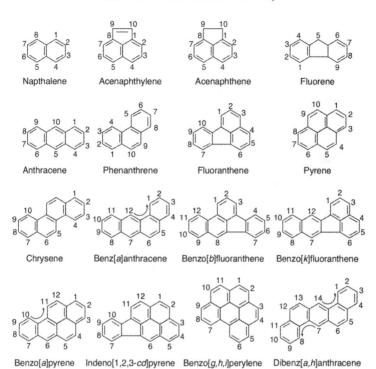

FIG 5.4. Structure of some polycyclic aromatic hydrocarbons (PAHs). *Credit:* US EPA.

tioned as a possible source of DIBs. The DIBs are found outside dense clouds, not in denser clouds, where stars and planets are formed. Because of carbon atoms' tendency to combine with other atoms, as explained in chapter 2, the most abundant interstellar molecules contain carbon. By this writing, two of the most massive interstellar molecules other than C_{60} and C_{70} are $C_{10}H_7CN$ (cyanophthalene) and $HC_{11}N$ (cyanopentaacetylene).

We have talked about carbon in interstellar space; let's turn now to its presence and role in planets, not only in the solar system but in exoplanets orbiting other stars.[5.4] As we've seen, all planets contain carbon, and their formation is a natural

FIG 5.5. Rendering of a protostar, a forming star surrounded by a disk of gas and dust. *Credit:* © Gemini Observatory/Lynette Cook/Science Photo Library.

consequence of star formation from material in interstellar dense clouds of gas and dust. Star formation starts with material collapsing to a disk, defining a plane where planets can grow surrounding the star. Many of these protostars, as astronomers call them, have been observed by the Hubble Space Telescope, often with "jets" of high-temperature, ionized gas perpendicular to the disk center. The jets and associated disk winds play some role in carrying away angular momentum and letting the disk evolve to form planets (figure 5.5).

Planets are, of course, one of the most interesting components of our Galaxy. Stars typically have multiple planets, so the Milky Way has many more planets than stars. Because they are so small, exoplanets orbiting stars are exceedingly difficult to detect and study. It is only because of revolutionary break-throughs in technology and data processing that it is possible to study thousands of exoplanets. Just a few decades ago, some doubted we could ever study exoplanets, and some wondered if planets even existed around other stars.

In 1584, Italian philosopher Giordano Bruno, a contemporary of Galileo, suggested that planets must encircle other stars, al-though they were unobservable. He also proposed life on these planets, similar to earthly life. These hypotheses were not offi-cially accepted by the Catholic Church, and he was burned at the stake in 1600 in Rome (mostly for other reasons, but his astronomical ideas didn't help). Starting in the later nineteenth century, a few claimed detections of exoplanets were put for-ward, based on stars' perceived cyclic motions, as though an unseen object was orbiting the star. The most prominent of these claims were by Peter van de Kamp and colleagues, who thought they were seeing tiny, periodic changes in the position of Barnard's Star.[5] They even deduced the unseen planet's mass. But later analyses of the same data, as well as observations made by different astronomers and telescopes, saw only statistical noise, not periodic changes. No planet.

Draugr, Poltergeist, and Phobetor are the names of the first observed exoplanets, discovered in 1992 orbiting a town-sized neutron star spinning 167 times a second. These pulsar planets constitute a quite unusual system that may have formed by the merger of two white dwarf stars to form a neutron star.

The first exoplanet orbiting a normal star was discovered in 1995 by Swiss astronomers Michel Mayor and Didier Queloz, for which they were awarded the Nobel Prize in Physics in 2019. They observed that the velocity of the star 51 Pegasi changed more than 100 meters per second every four days in a repeat-ing rhythmic manner. The speed of the star, measured by the

Doppler shift of its spectral lines, changed because the star was actually orbiting around the common center of mass between it and a planet with half the mass of Jupiter. The discovery was confirmed in a few days by Geoff Marcy and Paul Butler, using data from their ongoing program at Lick Observatory. Just a few weeks later they announced the discovery of planets orbiting two other stars. The floodgate was opened, and many more exoplanets were found by this Doppler velocity method.

When an exoplanet crosses the line of sight to a star, it causes a tiny dip in the star's brightness. For example, if Earth crossed in front of the Sun for a distant observer, the Sun would dim by 0.01 percent for a few hours. If the observer was far outside the solar system, this transit would repeat every year. The depth of the dip is a way to determine the size of the planet (figure 5.6). Sometimes its mass and atmospheric composition can also be derived. If the planet has an atmosphere, the transit provides a chance to explore its "air." The first successful detection of a transit was made in 1999. Transits were convincing proof that exoplanets discovered by the velocity method were in fact genuine planets and not false signals due to star spots or other stellar attributes, as some had suggested.

When exoplanets transit in front of their stars and block part of the starlight, the resulting spectrum contains features from the star as well as faint signatures from the planet's atmosphere. One goal of transit observations is to detect oxygen or methane in the atmospheres of Earth-like exoplanets. This is difficult, though not impossible. In the infrared part of the spectrum there are several spectral bands of water vapor, oxygen, ozone, and methane. On Earth, the main source of atmospheric oxygen on land is plants and in the sea, phytoplankton. The detection of oxygen in the atmosphere of an exoplanet could be a sign of the presence of life like our own. The search for such "biosignatures"

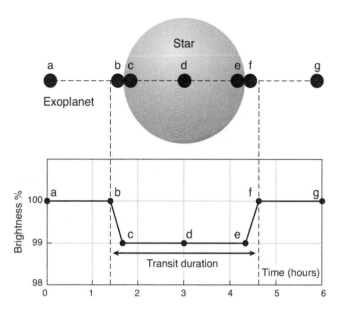

FIG 5.6. Exoplanet transiting disk of a star, showing the stages of brightness change as it crosses. *Credit:* © Institute of Physics and IOP Publishing.

in the atmospheres of Earth-like exoplanets is a major goal of the giant 6.5-meter James Webb Space Telescope put into space at the end of 2021. A great biosignature with a strong likelihood of being produced by life would be something that could not easily have been made by nonbiological processes. A good example would be an atmosphere out of chemical equilibrium, for example one that contained both free oxygen and methane. Earth's atmosphere, a mix of nitrogen, water vapor, and oxygen is certainly not a product of nonbiological chemical processes.

Methane might be a sign of life, and it has actually been detected in the atmosphere of an exoplanet. But the planet is Jupiter-like and very close to its star, a hostile environment for life as we know it. Nonbiological methane is present in all of the solar system's giant planets, as well as on Pluto and one of Saturn's moons.

Two spacecraft have been launched to find and analyze exoplanets that transit in front of other stars. Both have been fabulously successful. The *Kepler* spacecraft initially monitored just one small patch of the sky for four years, and then other regions for an additional three years.[5.5] The first exoplanets discovered were close to their stars, because they have very short orbital periods and a higher probability of passing in front of their star. Many orbit in days, not months or years. The first exoplanet discovered using the Doppler method orbited in just 4.2 days, and its orbital radius is only 5 percent of the distance between Earth and the Sun.

The Transiting Exoplanet Survey Satellite (TESS) followed *Kepler*. TESS surveys most of the sky, and its specialty is the study of bright stars closer to the Sun than those monitored by *Kepler*. TESS has found more than two thousand exoplanet candidates. In mid-2023, the total number of exoplanets by all methods had reached 5,500 (figure 5.7).

Many stars were found to have two or more planets, and some are in the habitable zone. The most common planets are "super Earths," an unexpected planet type that is somewhat more massive than Earth and not found in the solar system. Some exoplanets that are just moderately larger than Earth have extended hydrogen atmospheres. This quite unexpected finding suggests that terrestrial planets might have commonly had hydrogen in their atmospheres, at least for limited periods when they were young. This phenomenon has interesting implications for the early atmospheres of Earth-like planets, even possible chemical processes of hydrogen in the mantle forming water.[5.6]

Studies of exoplanets indicate that nature produces an amazing range of planets. Many of these differ from what we have in

Gas Giant
Similar to Jupiter or Saturn

Neptune-Like
Similar to Uranus or Neptune

Super Earth
More massive than Earth,
lighter than Neptune

Terrestrial
Rocky, in Earth's
size range

FIG 5.7. In addition to their various types of orbits, there are at least four distinct classes of exoplanets: gas giants, Jupiter-like planets; ice giants, like Neptune; rocky terrestrial planets; and super-Earth planets, unlike any planet in our solar system and intermediate in mass between Earth and Neptune. *Credit*: NASA.

the solar system, and they also have a range of evolutionary histories. Aside from their physical properties, the presence of planets with elliptical, close-in, far-out, and even retrograde orbits is strikingly at odds from what was expected for other planetary systems. As new telescopes allow better study of Earth-like solid planets, we may have more surprises. For example, it is possible that planets formed around carbon-rich stars, those with more carbon than oxygen, could be quite different from ours. Rocky planets around these stars could feature exotic mixes of minerals such as silicon carbide, graphite, and diamond in their interiors.[5.7]

One of the astounding findings from the studies of exoplanets is that we now know the value of F_p, the fraction of stars that have planets. F_p is one of the seven factors of the Drake equation that, when multiplied together, provide an estimate of the number of active, communicative extraterrestrial civilizations in the Milky Way. This equation was first promoted in 1961 by radio astronomer Frank Drake.[5.8, 5.9] F_p could have been anything between zero and one. A value near zero would be devastating for searches for radio signals from alien civilizations. Zero times any other number is, of course, zero. The abundance of exoplanets that have been observed, proves that F_p is close to one. It is not a barrier for the search for aliens, carbon based or otherwise, that might be sending us radio or even optical messages across the abyss of space.

CHAPTER 6

What Is Carbon Good For?

We have discussed many aspects of carbon in the Milky Way Galaxy, but now we get up close and personal. What is the use of this element in our lives? Well, to start with, we couldn't possibly be here without it. It is indeed very true that "we are billion year old carbon." We are made of the sixth element; we live in it; we are surrounded by it; we eat it; we exhale it; we think with it; the list is just endless. Carbon is essential in almost any product from the car you drive to the bed you sleep on, to oil paintings to the smartphone screens that modern people have come to depend on. In this chapter, we will discuss just a few selected uses of this amazing element.

Fossil Fuel

Our fossil fuels were made in the distant past. Much of our coal formed from plants that grew during the Carboniferous Period, but our oil and natural gas deposits were largely formed from marine organisms that lived during the Cenozoic and Mesozoic periods. Probably our primary use of carbon is still producing energy, in a wide array of forms and uses. Energy from burning carbon powers most of our cars, aircraft, and ships, heats our

TABLE 6.1. Total U.S. energy production, 2020

Petroleum	35%
Natural gas	34%
Coal	10%
Renewable	12%
Nuclear	9%

Source: US Energy Information Administration

Note: Renewable sources include biomass, hydroelectric, geothermal, wind, and solar energy.

houses, cooks our food, and helps make the fertilizer needed to support the human food supply. It also makes our clothes and houses and stores our electronic information. Producing all of this energy also drives global warming. At the present time, carbon-based energy sources still dominate the world's source of energy, as they have done for millennia.

Oil has played a key role in the development of our modern world. In 1857, Edwin Drake settled in Titusville, a small town in western Pennsylvania. He had heard that oil was seen seeping out of the ground on a nearby farm, but it was only in puddles, not in useful quantities. The benefits of oil have been known for millennia. Though cars hadn't been invented yet, oil lubricated the engines of the Industrial Revolution and all the moving joints that made up machines, from wagon wheels to locomotives. Oil was also highly sought after for its use in kerosene lamps, a major source of lighting at the time. Drake immediately tried to find a way to extract the oil.

Drake, along with George Bissell, a New York banker, transferred a drilling technique used for salt mining to oil drilling. In 1859, after several frustrating dry holes, the two found oil in great enough quantities to be cost effective. Immediately, an oil rush began, and wells spouted around the region. At the time, about half of the world's oil production was centered in western Pennsylvania!

FIG 6.1. In 1901, this Spindletop gusher produced 100,000 barrels a day and started the modern petroleum industry. *Credit:* Photograph by John Trost.

Years later, huge discoveries were made in Texas, California, Oklahoma, and other locations, and the US oil business has never been the same. The first major discovery, named the Spindletop, was in Texas, and it dwarfed all the oil production in Pennsylvania (figure 6.1). The discovery of Spindletop is the stuff of legend. The area where the source was discovered was a salt dome, hence the name. After some years and funding now and then, in 1901 a group found oil at a depth of about 1,140 feet. The resulting gusher extended to a height of 150 feet and produced a massive flow of oil, the likes of which had never been seen before.

Within months, the oil industry moved from Pennsylvania to Texas. Naturally, a criminal element entered the picture, as with any discovery of valuable resources. In this case, slanted wells were dug next to legitimate, registered wells, to steal oil. The Texas Rangers had to come in to settle things. In parallel, oil was being discovered elsewhere around the world, in the Middle East, Canada, Brazil, Kazakhstan, Russia, Norway, East Africa, and Mexico, which has a huge reserve. And there's a lot more under the world's seas, near the perimeters of continents, but oil from under the seabed is difficult and expensive to recover, and it sometimes leads to massive problems. Among famous oil disasters are the 1979 Ixtoc oil spill in Mexico, the 1989 Exxon Valdez tanker wreck, and the 2010 Deepwater Horizon oil spill in the Gulf of Mexico.[1]

Our increasingly efficient use of gasoline consumption as well as the introduction of hybrids and electric cars reduces our dependence on oil, but not yet by much in the big picture. Ultimately, sustainable sources of energy will become a necessity because fossil fuels will run out, be banned, or become too expensive for common use. At present, however, the world has an expanding population that still has an alarming desire for fossil fuels. Rising costs and the utilization of new extraction methods such as hydraulic fracturing (fracking) are expanding capabilities for recovering oil and natural gas from vast deposits of natural gas and shale oil deposits. All of these efforts have broad-ranging economic and environmental impacts.

Another source of energy is coal. One can only wonder what our modern world would be like if we did not have access to the vast amounts of energy from coal, nature's gift to us from the Carboniferous Period, over 300 million years ago (figure 6.2). Most of our planet's treasure of uniquely burnable rocks was made during a period of Earth history when vast wetlands were buried and capped by sediments, and their organic material turned into coal instead of just rotting away. It seems ironic that

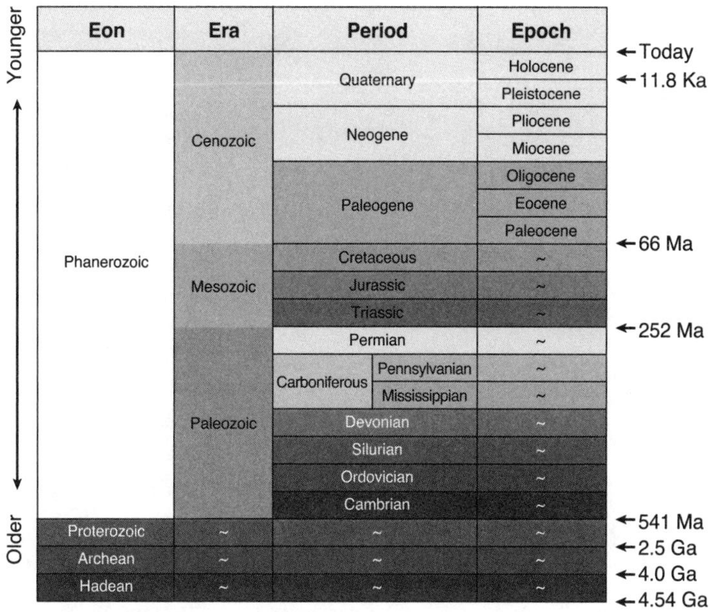

FIG 6.2. Geologic time. *Credit:* Image by Jonathan R. Hendricks for the Earth@Home project.

much of our coal was made in a period of history when CO_2 levels varied wildly from much higher to much lower than the present concentration. The low levels are thought to have resulted from rapid plant growth and burial. Major lowering can lead to entering a glacial period or worse.

The innumerable benefits of coal go hand in hand with long-known problems that include both air pollution and just getting it from the ground. Coal mines scar Earth's surface, produce dangerous waste products, and cause fatal accidents (figure 6.3). Coal is a 100 percent natural organic product, but burning it emits famously unhealthy gases and particles. It also, of course, generates great amounts of the greenhouse gas carbon dioxide, much more so than burning natural gas. The CO_2 emissions from electricity

FIG 6.3. Australian coal mine. *Credit:* EMBER/Sandbag Climate Campaign.

generation by coal are twice that from natural gas. Coal mining also presently releases more of the potent greenhouse gas methane than either the gas or oil industries. It is sobering to realize that coal is the most abundant known supply of fossil fuel that could be used to power future generations.

There are several grades of coal, from low-grade soft brownish lignite to high-grade, very hard black anthracite coal. The coal used in power stations is normally bituminous coal, midway on the hardness scale. Until about fifty years ago, subbituminous coal, just up from lignite, was mined where one of us (TPS) lives, near Boulder. This grade of coal is so flammable that it was transported by train in the winter, when the outside temperature is low. These days, a more stable bituminous coal is carried by train, mostly from west of the Rockies.

There are two general methods of getting coal out of the ground, open-pit and underground mining. Underground is the most common (60 percent of the total, in the US)—and the most dangerous. Pillars made of coal supporting mine ceilings can collapse, and escaping gas can poison miners.[2] Black lung disease is another notorious coal-mining danger. After centuries of growth, coal mining is anticipated to begin a significant global decline due to environmental concerns. Some believe, however, that the process of carbon sequestration could prevent this and keep coal mining active long into the future. One proposed approach for carbon capture and storage (CCS) is to inject coal-combustion products into deep reservoirs such as high-pressure saline aquifers. Technically this is quite possible and is essentially a case of reverse mining, although it would consume a significant fraction of the energy produced by burning coal. Coal would be mined, burned to produce energy, and the carbon dioxide generated would be isolated from the atmosphere by locking it up in long-term geological storage.

The US has the largest known recoverable coal reserves, followed by Russia and Australia. The leaders in known natural gas reserves are Russia, Iran, Qatar, and the United States. Natural gas can't be used as fuel until contaminants (e.g., heavier carbon-bearing molecules) are removed. Natural gas is considered to be the cleanest fossil fuel, but it does have downsides in addition to producing greenhouse gas. This flammable gas causes fires, explosions, and asphyxiations, though only rarely. As energy sources go, it is considered safe, except during earthquakes or other events that rupture gas lines.

The buildup of atmospheric greenhouse gas from fossil fuels is driving the transition to green energy, which ideally means no carbon burning. Energy sources such as hydroelectric, geo-thermal, solar, and wind farms are increasingly available and being used, but we won't be free of our reliance on fossil fuel–based energy for quite a while. Ultimately, our supply of black rocks, oil, and natural gas for burning will be exhausted, and sustainable energy sources will be our only source of power. The energy transition, our evolution past fossil fuels, is occurring, and its economic, technical, and societal issues are some of the great challenges of our time. It is interesting to ponder the evolutionary energy sequence from firewood, to coal, to oil, to gas, and finally to electricity from dams, solar cells, wind turbines, or nuclear reactors. Except for nuclear, all of this energy came from the Sun.

One alternative to fossil energy sources is nuclear power. Around the world, about 10 percent of all electrical power is generated by nuclear reactors. France gets more than 70 percent of its power from fission reactors, and China gets about 5 percent but plans to build 150 new power reactors by 2035. A number of countries, including Australia, have no nuclear power. The US is the world's largest nuclear power producer, and reactors have provided nearly 20 percent of our electrical power over the past thirty years.

Nuclear power has some advantages: a vast supply of energy; ideally no (or not much) pollution of the air or water resources; and not much mining to acquire uranium—and even that is decreasing with the use of breeder reactors. While nuclear energy is promoted by some as green energy, the nuclear reactor core meltdowns and infamous disasters at Chernobyl in 1986 and Fukushima in 2011 are clear lessons of the environmental risks of nuclear power and the uncertainties of our ability to adequately understand those risks. Following Fukushima, Germany began a program to phase out its nuclear power program and the use of coal. Germany's 2022 goal was to provide 100 percent of its total energy from renewable sources by 2035. But the impacts of this transition are continued subjects of discussion, and the situation is complicated by the war in Ukraine and multiple facets of global warming.

Gradually, some people are realizing the benefits of nuclear reactors over mines and oil wells. Like many things in life, it's a trade-off. On one hand, you have radioactive waste that will last for millennia and materials that could be used in bombs; on the other hand, you prevent the dangers of large-scale mining or drilling and reduce carbon ejection into the atmosphere.

A way to get around these problems might be the use of fusion reactors. Instead of splitting heavy atomic nuclei, here hydrogen nuclei combine, creating energy in the process. Stars do it; why can't we? If fusion reactors become practical, they would be an evolution from our long dependence on energy from carbon to getting energy from fuel in our oceans. Fusion reactors are fueled by hydrogen, and the ocean holds impressive supplies of H_2O. Fusion reactors, ideally, don't produce waste products, and we have an inexhaustible of supply of fuel. But there is a hitch: the difficulty of getting two atomic nuclei close enough together to fuse. They have positive electrical charges that repel each other. The nuclei have to be brought very close together and that takes extremely high temperatures. In stars, the pressure needed to support the weight of upper layers is obtained by core temperatures that are high enough for fusion to occur. So we have to create conditions that emulate stellar interiors, and that's difficult, to say the least. Several projects around the world are developing methods to develop fusion reactors, but the cost and technical difficulties make for slow progress. For the past half century, the breakthrough to fusion has been our carrot on a stick, just out of reach, always just a decade away. As Kermit the Frog used to say, "It's not easy being green."

Light

Living on a spinning globe illuminated by sunlight, we spend half of our lives on our planet's dark side. The ability to light up the darkness was a profound human invention that, like making fire, distinguished us from all other vertebrates. Archaeological evidence shows that humans were making oil lamps over ten thousand years ago, burning animal fat or oil from olives or nuts. From the 1700s to the mid-1800s, the preferred oil in much of the world was obtained by heating whale blubber in large metal pots. As whaling declined, lighting switched to kerosene, a refined hydrocarbon made from either petroleum or oil made by heating coal. After the Civil War, kerosene lamps were widely used to light up homes, trains, ships, and other places where people gathered. In commercial applications like theaters, businesses, and city streets, gas lights were used with the gas coming from either natural gas or gas produced by large facilities that roasted coal to produce coal gas. Lights for the first cars used acetylene (C_2H_2) gas produced by a mixture of calcium carbide and water.

Lighting began a fundamental change in 1880, when Thomas Edison and his workers developed a revolutionary electric light that could radiate light for over 1,200 hours (three hours a night for a year) before burning out. This first practical light used a filament made from a special bamboo from Japan. When heated, the filament carbonized to nearly pure carbon. Bamboo filaments were used for a quarter of a century before being replaced with tungsten wire that made whiter and longer-lasting light. Now, after thousands of years of using carbon-based lights, the world is shifting to light-emitting diodes (LEDs). These semiconductor lights are more efficient and have long lifetimes. The semiconductor junction that emits light does not contain carbon, but many LEDs contain polycarbonate plastic as part of their structures. It is possible that future LEDs may utilize new forms of carbon. Carbon dots, for example, are a nanomaterial only 10 nm in size, and their unusual luminescent properties are being explored for use in LEDs.

Transportation

The Industrial Revolution changed everything to do with transportation. Before the mid-nineteenth century, transportation relied on wind, steam, and animals. Currently, most modes of transportation are fueled by vast amounts of carbon-based compounds. Gasoline, jet fuel, and coal have powered the majority of our cars, trucks, airplanes, and ships. Without carbon, it is hard to get anywhere.

Energy from carbon has been critically important for transportation, but an often-unappreciated role of carbon-rich compounds has been their role in mechanical lubrication. Petroleum and other slippery, carbon-rich fluids have made it possible to build essential moving parts. Chariots, covered wagons, steam engines, internal combustion engines, gear boxes, and turbines were only possible because of the role of thin films of carbon-based lubricants.

A revolutionary development made possible by carbon was the invention of the car. The first horseless carriage was invented in 1885 by Carl Benz. This was a vehicle with a simple engine, one cylinder, three wheels, and a set of chains to transport energy from the gasoline-powered engine to the wheels. In an early public demonstration, Benz managed to hit a rock wall—the steering mechanism was a tiller with no fine control. Despite that setback, Benz went on to sell his automobiles, reaching more than five hundred vehicles sold in one year (1899).

Benz (and helpers) manufactured cars one by one at first. In 1902, Ransom Olds developed the concept of a production line, later improved and expanded by Henry Ford, he of the Ford line of cars and trucks (figure 6.4). In 1913, moving assembly lines reduced the time to make a Model T from half a day to an hour and a half. Ford built huge factories where raw materials such as iron ore and coal rolled in and finished cars rolled out. As a result, cars became inexpensive enough that they were readily available to the average person. All of this depended on carbon. The cars and engines were made of steel strengthened with carbon, powered with gasoline, lubricated with oil, rolled on carbon-based tires, and painted with Henry Ford's famous black carbon-based paint. The car led to the building of over 2 million miles of carbon-rich asphalt roads in the US alone.

Ford was an innovator in several ways. Not only did he advance the production line concept, he found that the parts of cars didn't have to be made in his plant; he purchased them from many sources. The Model T, built up that way, dominated the transportation market for two decades.

Nearly 100 million cars are built each year, and the technology is rapidly advancing. The only electronics in cars used to be the radio and lights, but modern cars are filled with computer chips that control the engine, give us directions, run the brakes, detect nearby cars with radar, warn when a collision is imminent, initiate air bags, and tell us when we have engine problems or reduced pressure in tires. Electric cars (EVs) typically have battery packs that weigh between 1,000 and 2,000 pounds and

FIG 6.4. Early car from Benz and a 1925 Ford Model T. *Credit:* Model T by ModelTMitch via Wikimedia Commons (CC-BY-SA 4.0).

power essentially everything in the car in addition to the motor. Many people believe that the family car will soon be fully "self-driving" and will allow drivers to "ignore" the road and perhaps take a nap if they wish. Some imagine that all cars of the ideal future world might be totally robotic taxis that move people and cargo anywhere they wish to go. Flying cars were well imagined back in the 1950s and some were even built, but it now seems quite unlikely that the anticipated George Jetson era will ever become a reality.

Iron and Steel

A use of carbon that must not be overlooked is in making iron and steel. Production of steel began about four thousand years ago, at the beginning of the Iron Age, crafted by artisans long before there was any semblance of a scientific approach to extending technology. Making steel for swords, knives, and strong tools evolved over time with the development of a considerable amount of myth and folklore on how it should be done to make the very best final product.

Steel is an iron metal alloy that contains small amounts of carbon and other elements that drastically improve its strength and physical properties. Iron, element number 26 in the periodic table, is a material that we use for buildings, bridges, railroad rails, ships, bolts, and things that have to be strong, durable, and sometimes very sharp. Extracting iron metal from iron ore (iron oxide) requires a prodigious amount of energy that is normally obtained by burning fossil fuel, and it requires carbon to chemically react with iron ore. Extracting a single iron atom from iron ore requires the chemical intervention of more than one carbon atom. Oxygen bonded to iron in iron ore reacts with carbon to form carbon monoxide or carbon dioxide, gases that escape, leaving metallic iron behind. The first iron was smelted with wood charcoal, but industrial production of quality iron required the high temperatures and energy output obtained by burning coal. This is normally done by using heat-processed coal that is turned into a purer form of carbon called "coke." The great abundance of coal found in many locations around the world fueled the phenomenal use of iron and steel in the Industrial Revolution.

The strength, durability, and formability of steel have played a critical role in our ascent from living in caves to the modern world. To turn iron metal into steel, you need to add the right amount of carbon. To become steel, iron needs 0.2 to 2 percent carbon. Less carbon creates wrought iron, used nowadays in decorative ways, like fences, signs, and lampposts. If 2 to 4 percent carbon is introduced, it becomes cast iron, which is brittle but has many valued properties, and can be cast into molds. It has been widely used since its invention in China over two thousand years ago.

Cement

Most of the world's near-surface carbon is not contained in fossil fuels, atmospheric gases, or in living things. It is locked up in carbonate, a rock that has been extraordinarily important to us because it is used to make cement, an invaluable building material since Roman times. Limestone (calcium carbonate) is heated to over 1,400°C, at which point the carbon has been driven off, leaving mainly white calcium oxide (lime). Lime is mixed with other materials to form cement that, when mixed with sand, rocks, and water, forms concrete, literally the foundation of our modern world. Within the next decade, the cumulative mass of concrete made by humans will exceed the living biomass of our planet. Without concrete, reinforced with steel bars, our world would be a radically different place. Concrete and iron, although largely taken for granted, are the miracle materials that make modern life possible. They go together because concrete beams and columns must have steel rebar inside them to provide the strength needed for modern construction and survival of stresses from earthquakes. The production of cement is responsible for about 5 percent of the world's production of CO_2. About half of this greenhouse gas comes from burning fossil fuel to heat limestone, and half is released from the chemical conversion of calcium carbonate to calcium oxide. Additional energy is also needed to mix the heavy materials and transport them to installation sites.

Plastics

Plastics come in many different forms, from brittle to very soft, from easily breakable plastic knives and forks to rubber bands. "Plastic" refers to many different forms of malleable or hard substances, all based on hydrocarbons. The degree of hardness depends on the molecular bonds between polymers, which are chains of repeating structures. Some plastics are unchanging with temperature variations, while others (thermoplastics) can change shape with changing temperatures and be molded into different shapes.

A natural carbon-based substance, cellulose, is the basis for cellulosic plastics. Cellulose was first isolated from plants by French chemist Anselme Payen in 1838. Cellulose is a hard and indigestible substance used in paper and clothing, not to mention high-fiber diets.[3]

Metallurgist Alexander Parkes created the first human-made plastic in 1856 as a substitute for ivory, specifically in billiard balls.

The substance, named Parkesine at first, was displayed in the London International Exhibit in 1862. That name didn't live long.

In 1907, a synthetic plastic was created and patented by Leo Baekeland, who named his invention Bakelite. Bakelite is not widely used these days, but this formaldehyde resin plastic was used extensively in the last century, appearing, for example, in electrical insulators, sockets for vacuum tubes, and the wires above trolleys. The "electrical smell" of old electronics that contain vacuum tubes is often due to the presence of Bakelite. If you stick your head into the cockpit of an old World War II bomber, you can often still detect a distinct formaldehyde smell from its Bakelite, even after half a century of airing out. When all tubes were powered on, their filaments glowing, the heated Bakelite produced the strong characteristic smell of "hot electronics," the smell of a bygone age (figure 6.5).

Many uses of plastics are so well known that a list could go on for many pages. Almost everyone uses grocery bags, water bottles, and zippers (and ziplock bags), and think of the toy industry! Without plastics, childhood would be very different and wouldn't involve disposable diapers, sippy cups, or pacifiers. You can see things made of plastic everywhere you look. Clothes, TVs, dishes and "glasses," cups, house exterior walls, car bumpers, carbon fiber airplanes, and on and on.

Plastics made movies possible, and they also were the material that formed the records, audiotapes, videotapes, CDs, and DVDs that brought music and images to billions of people. Many of the storage media used for music, movies, software, and data were in the form of disks. As we increasingly move deeper into the digital era, where things are stored in the cloud, or at least on servers, these plastic disks are doomed to be discarded and forgotten, like the paper IBM punch cards that preceded them.

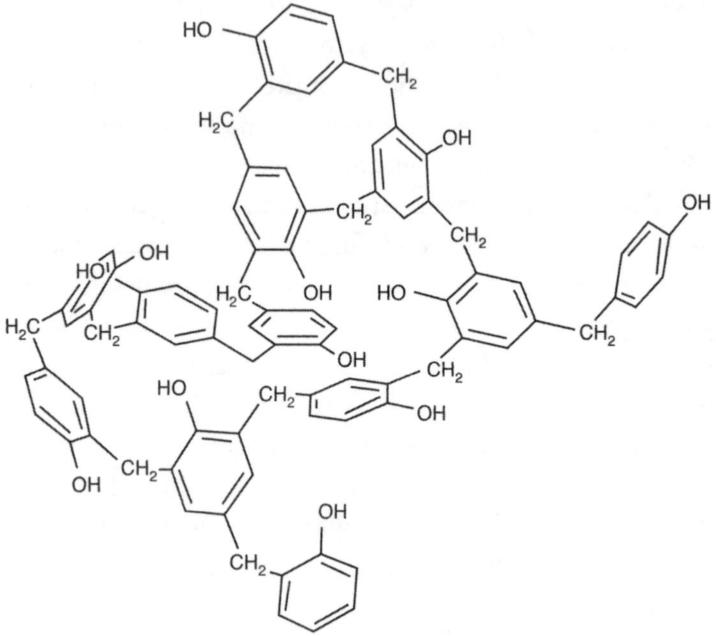

FIG 6.5. Bakelite structure. *Credit:* ChemSketch 8.0 via Wikimedia Commons.

But plastics such as amber, which is formed from tree sap, were made naturally on Earth millions of years ago. Plastics are likely to occur on other bodies as well. NASA has announced the discovery of propylene, a gas in the atmosphere of Titan, the giant moon of Saturn. Polymerized propylene is called polypropylene, a widely manufactured plastic used in containers (which can be identified by its recycling code, "5," on the bottom).

Tape

Without tape of one type or another, life would be messier than it already is. The first tapes were made from cloth or paper, but many tapes today are made of plastic. The invention of tape occurred in antiquity, long before it was used to wrap mummies. A major advance in the development of tape occurred in the 1920s, thanks to 3M, a company that made sandpaper. This new tape was used for painting cars to prevent paint of different colors from blending together, and it was called masking tape. "Duck tape" came later, in World War II, and it shed water like a duck. There seem to be two claims to its invention; the first was engineer Richard Drew at 3M, and the second was a subsidiary of Johnson & Johnson during World War II, which used it to prevent moisture from entering ammo cases. The two inventions seem to be independent of each other. The term "duct tape" was later used to promote its use as a means of sealing the joints in air ducts. Duct tape is specially modified to provide better qualities, such as heat resistance.

Transparent tape, called Scotch tape, was invented by the same Richard Drew in 1930. This cellophane tape lets light come through, a benefit for torn papers of any kind, from book pages to dollar bills. Post-it introduced a tape that sticks to everything but can also be easily removed.

Carbon is involved in the composition of any type of tape. Even metal tape still has an adhesive composed of polymers, long chains of repeating molecular units. A polymer is constructed with a backbone of carbon atoms, held together by covalent bonds. Teflon, too, a component of modern cookware, is made of carbon-based molecules. A plastic with many unique properties, Teflon has a longer history than most people know. Its technical name is polytetrafluoroethylene (PTFE). Like many chemical products, Teflon was discovered accidentally. In 1938, Roy Plunkett, a chemist working for a company called Kinetic Chemicals (later assimilated by DuPont), was experimenting with chemical coolants and found PTFE instead. The Teflon name was registered in 1945.[4] Meanwhile, Plunkett went back to research on coolants such as Freon and gasoline additives, also carbon-based compounds. In 1973, Plunkett was inducted into the Plastics Hall of Fame.

The opposite of slippery surfaces is Velcro, which is made of nylon fibers, again carbon based. It was also invented before most people realized: in 1941. That year, Swiss inventor George de Mestral, noticing how burrs stuck to his dog's fur, thought something like that might have a use. Velcro has two sides, hooks and latches. When the two sides meet, the hooks get embedded in a sea of latches, making it difficult to tear the two surfaces apart. The use of synthetic fibers provides a great advantage over burrs sticking to dog fur in that they can be used over and over again.

We all know Velcro because it is everywhere; it did turn out to be really useful in many applications, such as tents, belts and shoes, backpacks and briefcases, carpet and curtain attachments to floors and walls, diapers, and more. NASA applies Velcro everywhere: in weightlessness, objects tend to float, and Velcro is an easy way to counteract that. Probably the most fun you can get from Velcro is Velcro jumping. You wear a jacket covered with one side of the Velcro, and when you jump against a wall covered with the other side, you stick.

From Antifreeze to Cocktails

We can heat with carbon, using fires, gas heaters, and furnaces. We can also cool with carbon—for example, as a coolant in a car radiator. Antifreeze, aka ethylene glycol $(C_2H_6O_2))$, lowers the freezing point and raises the boiling point, so water stays in liquid form for a much wider range of temperatures than pure water.

People who drink have a fondness for C_2H_5OH, better known as booze. Its formal name is ethanol, but who cares about such formalities at a party? You can find books about alcohol, even whole shelves of them, so we won't say much more. But something you probably didn't know (if you are not an astronomer) is that alcohol exists in space, in interstellar clouds and in comets. Even though the space between stars is nearly a vacuum, space is large. One technical paper in a highly respected journal reports about 10^{28}, or ten thousand trillion trillion fifths, of booze in a single interstellar cloud.[6.1]

Rubber

To get back to the more practical uses of carbon, rubber, like plastic, is everywhere. This odd substance got its name because it can erase pencil marks on paper by rubbing. Apart from rubber balls, dog bones, and other playthings, there is rubber in tires, food containers, pencil erasers, cheap pillows, engine and refrigerator door gaskets, and engine belts. Rubber O-rings are used to make leakproof seals on devices ranging from shower valves to rocket engines.

Rubber was discovered as far back as 1600 BCE, when it was used in South America to make balls for games and religious ceremonies as well as bindings to hold ax blades to handles. Rubber came to the attention of Europeans in the mid-1700s, when French explorers brought home samples with magical properties from Brazil. From there, rubber plantations were founded in the East Indies, leaving Brazil out of a growing industry. Natural rubber, like latex, is tapped from rubber trees, much like maple syrup is harvested from maple trees.

Untreated rubber is not good for many things, because it gets brittle when cold and sticky when hot. An early US rubber boot industry went cold because of these defects. Enter Charles Goodyear. Upon hearing about rubber, he purchased a life preserver made of it. He thought of other uses for it, if he could find a way to preserve its properties at any temperature. He found that mixing rubber in a nitric acid solution would "cure" it so that it remained durable at low temperatures. After a few failures in finding financial support, he finally received a contract to manufacture rubber mailbags.

Goodyear's mailbag company failed because only the exterior of the bags was resistant to heat and cold. The bags would tear in bad weather. Goodyear kept plugging along, though, and finally found a way to solve the problems. By experimenting with different additives to rubber, he accidentally dropped a mixture of rubber and sulfur onto a hot stove. The result was a substance that held its shape at a wide range of temperatures. This process is called vulcanization, and most rubber products are now vulcanized. A prominent tire company is, of course, named for Goodyear.

Tires keep our world on the move, but what are rubber tires made of? Of course, they are made of synthetic rubber, but they are also heavily reinforced with cords made of polyester plastic, steel, or even carbon. Rubber tires also contain a less obvious but critical component called carbon black. These tiny smoke-like particles of pure carbon soot make up about a quarter of a tire's weight, and they play a major role in making tires durable. They also are the reason why tires are black.

World War II depleted the supply of natural rubber, so a synthetic version made from crude oil replaced natural rubber. At present, synthetic rubber holds about 75 percent of the market. During the war, the search for substitutes for natural rubber led to an unusual and quite entertaining carbon product called Silly Putty. Its unusual behavior includes bouncing, stretching, dripping, and making imprints of cartoons from the Sunday papers. It is made of carbon, silicon, and boron, and is traditionally stored and shipped in a plastic egg. Let's just say there wouldn't be Silly Putty without carbon.

Age-dating History

One of carbon's great gifts to us is that it provides us with a means to accurately determine the ages of events that occurred thousands or even tens of thousands of years ago, thus providing a way to explore human history since the dawn of civilization. This marvelous property is not from normal carbon but from the rare radioactive isotope ^{14}C, or carbon-14, that has two more neutrons than carbon's most abundant isotope ^{12}C. Carbon in a growing plant is dominated by the isotopes ^{12}C and ^{13}C, but about one in a trillion atoms are ^{14}C. Carbon-14 is made in the stratosphere, and it decays with a half-life of 5,730 years. This isotope was first produced and shown to exist in Berkeley, California, in 1940 and was first found in nature by analyzing Baltimore sewage.

Carbon-14 is made by cosmic ray bombardment into the atmosphere from space. Cosmic rays generate free neutrons that turn nitrogen into carbon-14 by the reaction

$$n + {}^{14}N \rightarrow {}^{14}C + {}^{1}H,$$

where n stands for a neutron and ^{1}H represents a proton (hydrogen nucleus).

After formation, ^{14}C combines with oxygen to form carbon dioxide, which is ingested by plants, some of which are eaten by animals. Even pure meat eaters, like lions, consume ^{14}C because their food chain includes vegetarians. When a plant or plant-eating animal dies, carbon stops being taken up, leaving the ^{14}C to decay. The development of ^{14}C as a proven dating tool was done by Willard Libby in 1946 at the University of Chicago. Proving that it was a legitimate method was done by dating objects with known ages, such as ancient Egyptian tombs and tree rings. Effects related to solar activity, nuclear bomb testing contamination, and ocean mixing could be taken into account, and this ultra-rare isotope produced a revolution in archaeology and provided insights into recent geological processes. An intriguing link in science history is that the scientific study of ^{14}C put into the atmosphere from atom bomb tests led to improved understanding of how atmospheric CO_2 cycles into the ocean and provided early evidence that burning fossil fuels was going to be a problem.[6.2] Between 1945 and 1963, nuclear reactions in atomic bomb tests nearly doubled the atmospheric ^{14}C, and it has taken sixty years for the "bomb spike" to be taken up by the ocean and allow the atmospheric ^{14}C concentration to return to its former level.

A very significant use of radiocarbon dating was its use in 2021 on "fossil" human footprints in sediment layers found at White Sands, New Mexico.[6.3] Dating seeds just above and below the footprint layers show that the footprints were made between twenty-one thousand and twenty-three thousand years ago. This dramatic discovery proved that America was peopled about ten thousand or more years before the widely recognized Clovis people came to North America and left caches of spectacular stone points. The dating of seeds containing carbon that is four half-lives old was possible because of orders of magnitude improvements in analysis methods. The original carbon work used quite elaborate means to measure the actual radioactive decay rate of samples. To detect even half of ^{14}C atoms in a sample, you would have to wait five thousand years for them to decay. Current work uses special mass spectrometers to actually count individual ^{14}C atoms before they decay. These large machines are called accelerator mass spectrometers because carbon atoms are accelerated to high energy to provide the means to measure tiny amounts of individual ^{14}C atoms embedded in vast numbers of other atoms. This provides the ultimate sensitivity and the ability to measure the ages of samples that are small and very old.

Another impressive utilization of carbon-14 is the record that it provides of ancient geomagnetic storms on Earth that were not noticed or recorded by humans. The most famous geomagnetic storm event that was noticed by people was the Carrington Event of 1859. A huge flare was observed on the Sun by amateur astronomers, and, less than a day later, high-speed charged particles reached Earth and produced dramatic auroras and strange magnetic effects; they also induced currents in telegraph lines that disrupted communication, caused sparking of lines, and even started fires. Some have estimated that an equivalent event today could cause extensive damage to communication networks, power grids, and satellites. A lesser solar flare in 1989 induced a geomagnetic storm that knocked out electrical power across large regions of Quebec.

In 2012, Japanese physicist Fusa Miyake discovered that she could detect older and larger radiation events by measuring carbon-14 in annual tree rings in very old Japanese cedars. She found an event in 774 CE that caused an increase of 1.2 percent in carbon-14 above the previous year.[6.4] Based on isotopic records, the 774 CE event was the largest in eleven thousand years, and it produced over 100 million radioactive carbon-14 atoms for each square centimeter of the planet. The event also produced radioactive isotopes of chlorine and beryllium found in ice cores in layers formed from snow that fell in 774 CE. The major events recorded in tree rings occur roughly once a millennium, and they are all called Miyake events. Other Miyake events occurred in 993 CE, 660 BCE, 5259 BCE, 5410 BCE, and 7176 BCE. Such energetic events coming from space are a serious worry in the modern world where most people depend on reliable electrical power as well as the ability to have high-quality digital information transfer everywhere at any time. The Miyake events definitely happened, but some are not similar to storms from solar flares, and their origin remains mysterious.

Miracle Fibers

Golf clubs, tennis rackets, bikes, sailboats, prosthetic feet, hockey sticks, kayaks, fishing poles, race cars, and space telescopes all have one thing in common. They all depend on graphite epoxy or graphite mixed with other polymers for their basic properties. They are made of graphite fiber–reinforced polymer, a material that is like fiberglass but uses carbon fibers instead of glass fibers. Carbon fiber is made by heating plastic fiber in an oxygen-free environment to the point where atoms other than carbon are driven off. The plastic is carbonized, something that also happens in the kitchen when you cook dinner on the stove for much too long. The fibers are very thin, typically less than 10 percent of the width of a human hair. A single fiber that weighs a gram is 10 kilometers long! Compared by weight, graphite epoxy is stronger than steel or almost anything else. Besides being strong and light, it has interesting properties, such as the directionality of its strength, and it can be fabricated in unusual ways. Parts for spacecraft are sometimes formed from sheets of carbon fibers that are impregnated with uncured epoxy material and simply cut with scissors and placed in an oven to be cured. A place where high-tech spacecraft parts are made out of carbon fiber composites can sometimes look more like Aunt Millie's sewing club than a traditional machine shop that turns blocks of aluminum into intricate parts.

Consider golf clubs. They evolved from hickory to steel and now, to carbon epoxy. The same evolution occurred in tennis rackets. Irate athletes can break both, but in golf, at least, it's easier to throw the misbehaving clubs into the nearest pond. The Boeing 787 Dreamliner is the first commercial airplane to use carbon fiber extensively, over 20 tons of it, which reduced its weight by 20 percent. Some of its giant parts are made on elaborate rotating jigs where fibers or tapes of fibers are wound onto forms and then baked to cure their epoxy binder.

Carbon epoxy is used for the frames of some racing bicycles, because it is strong and very light, and it has superlative properties also in boat construction: a boat hull made of carbon goes way faster. This applies to any boat, from canoes to racing sailboats. Every sailboat entered in the America's Cup race now has a hull of carbon epoxy.

How about cars? Carbon epoxy is lighter and stiffer than the traditional construction material for cars. While Lamborghinis and a few other luxury cars do contain carbon fiber parts, most cars do not because of their considerable expense. It is quite likely that carbon fibers will be increasingly incorporated into standard cars as wider use brings its cost down.

Besides fibers, there are other forms of carbon that will surely play important roles in the future. People are fascinated with buckyballs, those tiny soccer balls with sixty carbon atoms and such great potential in applications ranging from lubricants, solar cells, and hydrogen storage to numerous uses in medicine, such as delivering antitumor drugs and fighting allergies. Carbon nanotubes, which are analogous to rolled-up chicken wire, are also of great interest. These are tiny tubes, only a few atoms across. Single-wall carbon nanotubes are the strongest and stiffest material known.

Writing and Printing

You may not have thought much about the sixth element's low-tech role in making dark markings on light backgrounds. Pencils, inks, Xerox machines, and printers all use carbon to make dark markings on white paper.

We humans began to harness this property of carbon tens of thousands of years ago, using soot, burnt sticks, and charred bones to make drawings on the light-colored rock lining the walls and ceilings of caves. The role of the black element in forming long-lasting art and information has profoundly influenced our evolution as an intelligent species and gave rise to civilization as we know it. Almost everything that we know about human history in the last few thousand years is known because of records written in ink. All of recorded history, from cave paintings to Dickens, Chaucer, Darwin, Magellan, Galileo, da Vinci, Curie, Newton, Shakespeare, Verne, Hemingway, Salinger, Beethoven, Mozart, Picasso, Tchaikovsky, and Einstein was "recorded" with carbon. Carbon-based ink is surely one of the greatest innovations in history because it provided an effective way to pass information, art, musings, doctrines, and vital documents around the world and on to future generations.

Despite its name, India ink was invented in China, over four thousand years ago. It is composed of very fine carbon particles suspended in water and a binding agent. The early source of carbon in ink was lamp black, the soot from flames. Ink was a vast technological improvement over wax and clay tablets and figures scratched on stone. With the invention of the printing press, carbon-based ink spread words across the world, eventually to billions of people.

Until the digital age, most things that people learned in school were learned by reading black ink marks on paper and answering questions with carbon marks on paper. Typewriters are now extinct, but until 1980 the sound of steel typewriter hammers pounding on cloth ribbons and pushing ink into well-formed letters on paper was ubiquitous on our planet. Businesses, governments, newspapers, schools, hospitals, and armies in the mid-twentieth century could not have functioned without them. Ink used in typewriters and other devices may just look like soot, but it is actually an amazing material composed of very tiny clusters of carbon particles made of hexagonal rings of carbon atoms. This aromatic carbon form results in very strong absorption of all colors of light. In its pure form, carbon is black, much blacker than white paper, and blacker than almost anything else. The contrast effect on a white background is so strong that even a very thin layer of microscopic carbon particles can make a very black mark that can last for centuries. Modern ink used in inkjets is actually a nanomaterial; the smallest particles in its clusters of carbon particles, at around 10 nm across, are only about a ten thousandth of the width of a human hair.

While a simple invention like ink certainly changed the world, like another brilliant invention, the lightbulb, its importance is diminishing, and it will never again be as important a medium of communication as it was in the past. Both ink and incandescent light bulbs are being replaced by light-emitting semiconductors. Books, print, and libraries have provided the backbone of civilization, but things are rapidly changing as we pass deeper into the digital age. People of the future probably will not have physical libraries, and it is likely that eventually they will think of paper books as quaint relics from the distant past. Paper books with ink words can't be electronically searched, they do not contain electronic links, high-resolution images, or videos, and they also don't contain all those advertisements that Googles-of-the-future will deftly throw at us for things they want us to buy. Ink will likely still be here a few thousand years from now but will be used for a decreasing fraction of the words that people read. However, though ink use may soon fall along the wayside as the digital revolution marches on, carbon itself will not be left behind. Its importance to the Information Age is already well established by its presence in components such as electronic displays used on TVs, laptops, phones, watches, and cars. Exotic carbon forms such as nanotubes have potential for widespread use in future electronic devices.

Certainly, other futuristic carbon products will be created, some useful and some perhaps useless. Graphene, discussed in chapter 2, is another example of a form of pure carbon that has already found important applications. Potential uses of graphene include superconductors, integrated circuits, very small transistors, solar cells, light sensors in a wide range of wavelengths, and digital cameras, as well as cancer and COVID detection. An exciting application of graphene, and its close relative graphyne, is desalination, the process of making pure water from seawater. Graphyne isn't industrially produced; it is only theoretical, but its potential uses are likely to stimulate efforts to make it. Graphyne has double and triple bonds, and its structure is somewhat of a designer material that might be altered to produce important new electronic components.

In this chapter we discussed many forms of carbon and ways they affect our lives. In the next chapter we will highlight a singular, rare, and spectacular form of the sixth element.

CHAPTER 7

Diamonds

Diamonds have been prized throughout history as gemstones, and the business of finding, cutting, and selling these flashing carbon crystals has been pursued and improved over many millennia. In more recent times, the uses of diamonds in medicine and technology have only increased their value and presence in our lives. In this chapter we pay homage to a very special form of carbon—beautiful, durable, useful, sometimes scandalous, and always sought after.

There are credible references to diamonds from six thousand years ago. The first diamonds, or "special stones," were found by accident in alluvial washes on the western coasts of India. A 59 BCE Sanskrit text compiled a list of admirable diamond properties: hardness, brilliance, and dispersion of color. Later, Pliny the Elder, a prominent Roman, noted their ornamental value. But diamonds only became available in quantity with their discovery in South Africa in 1866, when a fifteen-year-old farm boy found a glittery stone that was later identified as a 21-carat diamond. Four years later, an 83.5-carat diamond was found, triggering a diamond rush that eventually attracted tens of thousands of hopefuls.

"THIS IS NOT THE TIME TO BE CARBON-NEUTRAL."

FIG 7.1. Lockhorns. *Credit:* © 2008 WM Horst Enterprises, distributed by King Features Syndicate, Inc.

There are four major volcanic pipes in the area near the town of Kimberley, South Africa. The first discovered was the "Big Hole" (figure 7.2). At 42 acres of surface area, this has been claimed to be the largest hand-dug excavation on Earth. The open-pit excavation ended near a depth of eight hundred feet, due to infill by debris and flooding by water, but underground mining eventually extended it to over three thousand feet. By 1914, 14 million carats of diamonds had been recovered from the Big Hole.

If you look at a collection of solid pure chemical elements, they are mostly a pretty monotonous bunch, but carbon is the superlative standout. A lump of pure carbon atoms can range from a black noncrystalline carbon to slippery crystalline graphite to crystal-clear diamond, which is the only room-temperature form of any solid element that can be transparent and "clear as crystal."

FIG 7.2. Kimberley diamond mine, known as the famous "Big Hole." The diamond pipe has been mined to a depth of over 1 kilometer. *Credit:* Photograph by Irene Yacobson. All rights reserved.

What differences there are between these different carbon forms! Noncrystalline carbon is glassy; it has no geometric order between its individual carbon atoms. Graphite, on the other hand, is perfectly crystalline. It is made of stacked sheets of carbon atoms arranged in hexagons linked together like chicken wire (graphene). The sheets are only loosely bound to each other, and graphite, the shedding dog of the periodic table, constantly sloughs tiny carbon dandruff flakes when it is handled. You can tell when a person handles a lot of graphite because their hands are dirty, covered by black, slippery graphite flakes. This gets on their clothes and anything that they touch.

There is no greater contrast to soft, sooty graphite than diamond, the hardest known material, which is clean and totally transparent in its pure form. You can use slippery graphite dust to ease your key in and out of a troublesome lock, but you would never try this with diamond dust. Its shards are gritty, used to abrade the hardest known materials. Diamond is extraordinary in many other ways; for example, it is also the best conductor of heat, has an extremely high melting point, and is a good electrical insulator.

Both graphite and diamond are forms of pure crystalline carbon, but graphite is composed of sheets, while diamond is composed of carbon atoms in a three-dimensional, strongly bonded "diamond structure" (figure 7.3). The unusual properties of diamond are related to its regular, repeating pattern of atoms: each carbon atom forms strong bonds with its four closest neighbors, with the angle between bonds being slightly more than 109 degrees. This diamond structure has cubic symmetry, and mined diamonds often display octahedral shapes—that is, the form of two pyramids joined at their bases. Out of the millions of compounds that carbon makes, diamond is the only one that features these bonds, which form a remarkably strong crystalline structure. These bonds are set in stone, so to speak, and they are responsible for the superlative properties of diamonds. In the previous chapter we saw many forms of carbon that are good for many things. The diamond, probably the most famous mineral, stands out from all other carbon forms in quite spectacular ways.

Although the statement that "diamonds are forever" is not always true, still, in most imaginable scenarios and conditions, diamonds can last nearly forever. Some diamonds in meteorites are actually older than the Earth and the Sun. In the right conditions, however, diamonds can turn into graphite or even into carbon dioxide. Under normal conditions, the conversion of diamond to graphite takes a tremendous amount of time, and

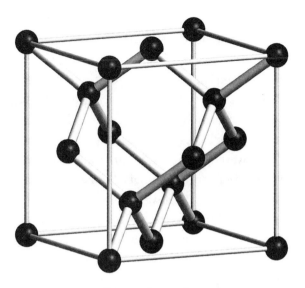

FIG 7.3. Diamond crystal structure.

there is no concern that this will happen to your diamond, or even the Hope Diamond, anytime soon. But heat a diamond enough, and it can turn black, actually becoming a fine polycrystalline mix of diamond and graphite. Black diamonds occur naturally, but sometimes clear faceted diamonds can be heat-treated to produce them as well. And when diamond is heated in air to extreme temperatures, it will burn into pure carbon dioxide and vanish. As previously described, the Lavoisiers burned up diamonds to confirm their theory that mass is conserved during combustion. This heroic and quite dramatic experiment, done with very large lenses to concentrate sunlight, was instrumental in showing that carbon is an element and illustrating what it means to be an element.

Although not as well known as diamonds, there is another form of transparent carbon called lonsdaleite. In its ultra-perfect

form, with no defects or impurities, lonsdaleite is predicted
to be even a little harder than diamond. Carbon atoms in lonsdaleite are arranged in a hexagonal structure, and lonsdaleite
is sometimes called "hexagonal diamond," but it is not actually
diamond, because it has a different crystal structure. The proven
sources of this extremely rare mineral are all associated with
either meteorites or large meteorite impact craters. Lonsdaleite
was first discovered, along with diamonds, inside meteorite
samples from the famous Barringer Meteor Crater near Flag-
staff, Arizona, and the greatest known source is the 100-km-
diameter Popigai impact crater in Russia, believed to contain
hundreds of thousands of tons of lonsdaleite and diamonds that
are mostly not of gem quality. (The Popigai diamonds were
formed from terrestrial graphite by the great pressures of the
impact event 35 million years ago.)

Special conditions are necessary to create diamond: in our
planet, the needed pressure and temperature are only found far
beneath Earth's surface. Most diamonds formed in the Earth's
upper mantle, at about 100 km below the surface of old continents. About 99 percent of diamonds that are mined are classified as Type I, and are believed to have been formed from carbon that descended from the surface, probably deriving from
organic compounds. Common diamonds have distinctive
properties, such as nitrogen content high enough to give the
diamonds a slightly yellow or grayish hue. Diamonds are nearly
pure carbon, but they do carry small amounts of other elements
and embedded solids. Only about 1 percent of small diamonds
have the color, clarity, and low enough nitrogen content to be
classified as precious Type IIa stones. Large diamonds are rare
and different in several ways, including their composition, age,
and birthplace. About half of the largest diamonds fall into the

FIG 7.4. The 1,109-carat Lesedi La Rona diamond discovered in Botswana in 2015, the largest gem diamond found in over a century. *Credit:* AP Photo/Seth Wenig.

elite IIa class; they have low nitrogen content and are exceptionally clear. Very large diamonds become famous and are often named. For example, the Lesedi La Rona diamond weighs in at 1,109 carats (figure 7.4).[1]

Over the past few years, a most interesting class of giant diamonds has been recognized. Almost all of these are classified as IIa. These monsters are clear, but they do contain tiny inclusions whose properties indicate that these best-of-the-best diamonds have a quite different history than common diamonds.[7.1, 7.2] They have superdeep origins—that is, they formed several times deeper than common diamonds. Superdeep diamonds that are large and have extraordinary clarity, color, and other properties are called CLIPPIR (Cullinan-like, Large, Inclusion-Poor, Pure, Irregular, and Resorbed) diamonds. The most famous CLIPPIR diamond is the Cullinan from South Africa, at 3,107 carats (0.61 kg!) the largest gem diamond ever found. It was cut into nine principal stones, all part of the Crown Jewels usually stored in the Tower of London. Not all superdeep diamonds are large; most are small, but they are scientifically quite important. Because small ones are more common, large numbers of them can be investigated to more broadly explore the processes and materials that exist in the depths where superdeep diamonds formed. Small superdeep diamonds are abundant in alluvial deposits in the Juina area of Brazil.

Superdeep diamonds are our deepest samples of solid matter from inside our planet. It is quite ironic that diamonds, long treasured for their beauty, are a previously unimagined and unique treasure of direct knowledge about the deep interior of our planet that otherwise would forever lie beyond our direct reach. We can send spacecraft to the edge of the solar system and even beyond, but it seems unlikely that we will ever build a device to retrieve samples from the top of the lower mantle at 660 km depth where the temperature is 1,600°C and the pressure is twenty-four gigapascals (240,000 atmospheres).

Superdeep diamonds are magic vehicles that transport tiny inclusions of deep materials to the surface, like small stowaways, protected by the incredible robustness of diamond. They provide a truly unique means to explore the Earth's deep interior, which includes the Mantle Transition Zone (MTZ) between 410 km and 660 km and even a bit below.[7.3] This zone marks the boundary between the upper and lower mantle, regions known to differ from each other because of their seismic properties.

Some superdeep diamonds contain inclusions of materials such as iron-nickel metal alloy, iron carbides, high-pressure minerals such as ringwoodite, and even exotic solids such as a high-pressure form of ice and gas containing hydrogen and methane. The presence of the ringwoodite in the MTZ is profound, because it can contain "water," which is actually hydrogen and oxygen bound as a hydroxide inside the mineral's crystal structure. Ringwoodite is a high-pressure silicate composed of Earth's most abundant atoms: silicon, magnesium, and oxygen. It has a similar chemical composition as the anhydrous mineral olivine (famous in Hawaii's green sand beaches) that dominates the upper mantle. Ringwoodite from superdeep diamonds has been observed to contain over 1 percent water, and so this mineral has the potential to hide an ocean or more of water near the MTZ.[7.4] (At depths below 660 km, ringwoodite transforms into phases such as bridgmanite, whose crystal structure cannot accommodate as much water.)

The high-pressure mineral ringwoodite was first discovered in 1969 as a purple silicate inside a meteorite that had been "shocked" in space by a collision with another solid object. In Earth's materials, ringwoodite has only been found in nature as small inclusions in deep diamonds. The discovery of ringwoodite in the MTZ has profound implications for the origin of our oceans and their long-term future. In the future, as the Sun becomes brighter, Earth will lose its oceans, but water deep in the mantle will continue to seep toward the surface, where it could form small ponds when our planet's surface temperature is below water's ambient pressure boiling point.

The differences between deep and shallow diamonds provide fundamental insight into the Earth's interior processes and history. The presence of metallic iron in superdeep diamonds, along with other evidence, implies that the carbon that formed these diamonds may have not been carried down from the surface, as with common diamonds, but up from below. It has been suggested that the carbon in deep diamonds was previously in iron metal. Such material is, of course, in Earth's iron core, but it also exists in the lower mantle.

Some superdeep diamonds also contain helium that has a different isotopic composition than what is found in the atmosphere, in rocks, and in upper-mantle diamonds. Helium has two isotopes, ^{3}He and ^{4}He, that vary with the number of neutrons they contain. Helium (He) in the Earth has two sources: (A) relatively young He from the radioactive decay of uranium and thorium inside Earth and (B) primordial 13.8-billion-year-old He atoms that formed in the Big Bang. The primordial component was carried to Earth in trace amounts inside the rocky bodies that accreted to form our planet. The biggest bodies were as large as Mars, and their impact could directly eject materials deep into our planet. Helium from radioactive decay in crustal rocks is nearly pure ^{4}He (only about one part per million ^{3}He), while the abundance of ^{3}He from the Big Bang, meteorites, and the Sun is on the order of a hundred times higher. Helium is an inert gas, and as the smallest atom, it tends to diffuse and mix between materials. Accordingly, it is remarkable that our planet retains at least two He reservoirs with different isotopic ratios. Whereas helium in a birthday balloon and in our planet's crust is nearly pure ^{4}He, some superdeep diamonds, as well as basalt from certain oceanic islands such as Hawaii, Iceland, and Baffin Island, are enriched in ^{3}He by factors of up to fifty. The isotopic signature of primordial helium is retained in deep materials. The data from diamonds play a key role in understanding what is going on, because their formation depth can be understood by analyzing the high-pressure minerals that they contain.

Diamonds form deep, but they travel to the surface in a most interesting way. In the Earth's past, there have been very unusual types of volcanic eruptions that magically lifted material to the surface from great depths. Material at depth, under great pressure, exploded upward, sometimes propelled to speeds of hundreds of kilometers an hour by expanding water and carbon dioxide gas. These strange eruptions formed long, thin "pipes," conduits only a few hundred meters across but a hundred kilometers long. When they erupted to the surface, some of the materials must have violently shot high into the air, but unlike other volcanoes, they did not deposit a mountain or other major amounts of material on the surface.

The up-rushing material ripped debris off the vertical walls of the pipes, and the shafts of the pipes were left filled with unusual rock types that are classified as either kimberlite or lamproite. The classic deposits are called kimberlite,[2] and they sometimes include a variety of rare high-pressure minerals, including diamonds (figure 7.5). Just to make things interesting for prospectors, miners, and investors, some pipes contain a fabulous wealth of diamonds and others contain none. Even the very richest deposits only contain about a carat (0.2 grams) of diamonds in a few tons of kimberlite, and most have much lower concentrations.

The pipes are often very hard to find because their surface features can be quite subtle. In the early 1990s, when fabulously diamond-rich kimberlite pipes were discovered in northern

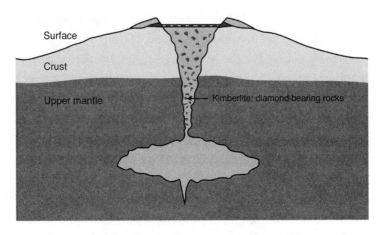

FIG 7.5. Diamond pipe cutaway drawing.

Canada, the stealthy pipes were first found as shallow ponds covered with ice. The ponds filled hollowed-out depressions that formed because the rocks in the pipe were softer and eroded a little faster than the surrounding rock.

Without the incredible volcanic diamond pipe process that elevates debris from great depths within the Earth, we would not know of diamonds except perhaps for the few that exist inside meteorites, rare bits of asteroids that have fallen to Earth. Without diamond pipes, the reasonably abundant diamonds in Earth's deep interior would never see the light of day. We would have no diamond rings or diamond saws, and we would never say that anything "sparkles like a diamond."

Because diamonds are so rare and potentially valuable, diamond-bearing locations are nearly always off-limits to weekend prospectors who might wish to find a small diamond. There is one place, however, where the average person can just walk in, dig around for a day, and have at least a fighting chance of finding a diamond: the famous Crater of Diamonds State Park near

FIG 7.6. Searching for diamonds at the Arkansas Crater of Diamonds State Park. *Credit:* © 2023 Arkansas Department of Parks, Heritage, and Tourism.

Murfreesboro, Arkansas, where visitors find around a thousand diamonds each year (figure 7.6). Although most of the diamonds are small (the largest found was 40 carats), some are of exquisite quality. They are found in the weathered surface exposure of an ancient volcanic pipe that is similar to the kimberlite pipes in South Africa but is composed of a rock called lamproite, a material that also fills diamond pipes in Australia. For a $13 entry fee, you can search the plowed surface of the pipe by screening buckets of soil, or, if you are really lucky, you might just spot a glistening diamond as you walk around.

In 2008, a tiny asteroid fell to Earth, breaking apart as it entered the atmosphere. Its fragments were fairly easy to find because they landed in a sandy desert in Sudan. Like some other meteorites, the fist-sized rocks contained small diamonds. Unlike the case with Popigai and Meteor Crater, these diamonds could not have formed on Earth, because the small rocks did not impact the ground at hypervelocity speed and generate shock pressures high enough to form diamonds. The diamonds were formed by the shock pressures of asteroids colliding in space at high velocity. Many meteorites have features caused by high-speed collisions, but most do not contain diamonds made this way.

There may be other sources of diamonds in the solar system. Both Jupiter and Saturn, and possibly Uranus and Neptune as well, have the right conditions in their atmospheres for "diamond rain"—plenty of carbon, and the pressure and temperature at some level in their extended atmospheres for gemstone-quality diamonds. And if diamonds are found in the solar system, they must also be in other planetary systems with the right environment for diamond formation. Thousands of extrasolar planets have been discovered, orbiting stars of all types.

A truly remarkable locale for the formation of diamonds is the dense interior of white dwarf stars, the end evolutionary state of most stars. Twenty years ago, it was found that the interior of the carbon-rich white dwarf BPM 37093 (nicknamed Lucy, after the Beatles song) is crystalline diamond! This was deduced from the pulsations of the star's brightness, caused by vibrations similar to the ringing of a bell. The emerging field of asteroseismology can use these observational data to infer interior properties, and it appears that white dwarfs are an important repository of diamonds in the Universe.

Just saying the word "diamond" is likely to generate a smile or at least a positive response from any person you are talking to. However, the story of diamonds on Earth is not all about glitter and light. The extreme value and attention given to diamonds have famously generated greed, thievery, cruelty, and tragedy over the ages. Diamonds inspire joy and awe, but they also have a dark side. A stone the size of a raisin can be worth a fortune, and the temptation to pilfer is something that a few people simply cannot resist. This has always been a problem with miners, because a miner can simply swallow a diamond and steal a stone that could have a value of millions of dollars, and there have been cases where miners were x-rayed every day in an attempt to prevent the smuggling of stones off the mine site. Diamond thefts are legendary subjects for books, movies, and news reports.

In addition to a history of theft, murder, and mayhem that surely dates back to the very first discovery of diamonds, these peculiar stones have often been imagined to have special powers. Besides being charmed stones with magical properties, some diamonds have also been associated with ill fortune. The most famous of these is the legendary curse of the spectacular blue Hope Diamond, now securely housed at the Smithsonian Natural History Museum.

Another prime example is the Koh-i-Noor diamond, which was found in antiquity and has an extraordinarily dark history of theft and contested ownership.[7.5] It was previously owned by numerous kings and emperors, and it is generally considered to be the most valuable gem on Earth. In 1850 it was presented to Queen Victoria, and it is now part of England's crown jewels. Even after a century and a half there are increasingly bitter feelings on which country should own this spectacular stone, which is only the size of a large piece of candy.

The Great Diamond Hoax of 1871–72 encapsulates the greed, ambition, and general nastiness that diamond fever can evoke. This was a time of great discoveries of gold and silver in the American West and in Africa, and a time when occasional diamonds were found in California placer gold deposits. It was also an era of great expectations, when enormous mining wealth was routinely pouring into San Francisco. During this period, there were numerous scams in which scoundrels sometimes salted their claims with gold dust to defraud investors. Philip Arnold and John Slack from Kentucky entered San Francisco with a bag of diamonds and a tale of a fabulous diamond strike (figure 7.7). With money from initial investors, they then traveled to England and bought $20,000 of uncut diamonds and rubies that they used to salt their "diamond strike" in western Colorado. Mining experts were led to the remote site, where they saw the gems on the ground and in anthills, confirming that the site was fabulously rich in diamonds. Investors eager to share in the riches poured money into the project. They included many of California's most successful bankers and businessmen, a former commander of the Union Army, a congressman, lawyers from both the Pacific and Atlantic coasts, and the founder of Tiffany & Co. Also involved at some level was Horace Greeley, whose articles in his New York newspaper stimulated the great westward migration. In this biggest hoax in the American West, the swindlers bilked investors out of $650,000.[3]

Fortunately, the fraud was exposed just before the investors lost millions and the city of San Francisco experienced a most embarrassing financial calamity, and it was a scientist who saved the day. The con was revealed to the keen eye of geologist Clarence King during a site visit. King noticed that gems were only found in places surrounded by footprints. Moreover, they appeared to have been placed in anthills, rather than being below the surface, and the surrounding rock was not consistent with diamond production. After saving the city and its easily duped investors, King became one of the most celebrated scientists of his time; in 1879 he was appointed the first director of the

THE DIAMOND FRAUD.

The Greatest Swindle Ever
Exposed in America.

HOW THE FIELDS WERE SALTED.

The Sharpest Men in California
Lose Nearly $2,000,000.

INDIGNATION IN SAN FRANCISCO.

Prominent New Yorkers out of Pocket
to the Extent of $750,000.

OFFICIAL REPORTS OF SCIENTIFIC MEN.

Disappearance of the Men who
First Found the Jewels.

FIG 7.7. The Great Diamond Hoax during the California Gold Rush. *Credit:* On the left is Philip Arnold (Wyoming History Society), and on the right is a newspaper story from the *New York Sun* in 1872 (*New York Sun*).

United States Geological Survey (USGS), and Kings Canyon and its National Park are named after him. The diamond-salted field is marked on current official United States Geological Survey (USGS) topo maps as "Diamond Field."

Surely the darkest side of diamonds in recent times has been the use of profits from rebel- and warlord-controlled diamond-mining activities to finance armed conflict and civil wars. Human rights violations by armed groups include the coercion of civilians, including children, to work long hours under deadly dangerous conditions. The billions in dollars of diamond-mining profits from these mines have fueled devastating wars that have killed millions in countries like Angola, the

Democratic Republic of the Congo, and Sierra Leone. The diamonds used to support these wars were termed "conflict" or "blood" diamonds in an effort to bring this dreadful activity to the world's attention. In 2002, the United Nations, along with diamond-trading nations and the diamond industry, developed the Kimberley Process to track diamonds and diminish the flow of blood diamonds into the legitimate world diamond trade. Greed and corruption are not just confined to diamonds. The mining of tantalum, tin, tungsten, and gold by armed groups has produced similar results to diamond mining, and their ores are sometimes called "conflict minerals."

After their recovery from mines, gem diamonds commonly start their journey into the commercial world with major auctions held by mining companies or their agents. The De Beers company is well known for its auctions, where selected buyers must pay for a diamond "lot," sight unseen. From auctions the gems are distributed to retailers—and then to you, with an enormous markup, maybe 300 to 500 percent or more. This is especially true for small, elite jewelry stores that have slow rates of sales, whereas the warehouse stores can have lower prices (and often lower qualities). Between the mine and the customer, gem diamonds often pass through an intricate web of dealers, a legacy of centuries of evolution of the international diamond business.

The famous De Beers company was founded in 1888 by Cecil Rhodes, and it has played a dominant role in the mining and distribution of diamonds for over a century. Rhodes was born in England but moved to South Africa as a teenager and eventually became the most important figure in the history of the gem diamond industry. The Rhodes estate funds the famous Rhodes Scholarship program, the most prized honor a college senior can attain, and Rhodesia, now known as the Republic of Zimbabwe, was named after him. Until relatively recent times, De Beers had a monopoly on the world's diamond business.

Once gem diamond rough is mined and sold, it needs to be cut to bring out its phenomenal beauty. Diamonds are the most difficult of all gems to cut, but because they are so desirable and valuable, techniques for cutting them have developed into an advanced art. The making of diamond jewelry, knowing how to strike in the right place to cleave the stones, was an industry in Europe by the 1300s. (This might have happened sooner, but an early tradition of austerity in Christianity held it back.) Before cutting diamonds was common, rough diamonds were worn by the upper class as status symbols or omens of good fortune. The first guild of diamond cutters formed in the fourteenth century in Germany, and the style of cutting spectacular symmetrical diamonds with diamond dust on rapidly spinning metal disks was developed in Antwerp in the fifteenth century. Today, most of the gem cutting is done in India; over 90 percent of the world's gem diamonds are cut in the Indian town of Surat.

Diamond may be harder than anything else, but it's actually fairly easy to break. When it's struck at the right place and at the right angle, it can fracture along cleavage lines. Diamond cutters use this property to initially shape chunks of a large diamond, and then fine shapes are made by grinding and polishing with diamond powder. Very large and costly gemstones can make for some nervous times. Imagine the stress of preparing to cut a diamond worth many thousands or even millions of dollars!

The reason that diamonds are cut is to bring out their fabulous sparkle and color, seemingly generated magically from a colorless stone. Cut diamonds sparkle because they are carefully crafted to reflect light to the viewer. They do this because of their shape and the unusual behavior of light inside them. Light travels only 40 percent as fast in a diamond as it does in air. In a list of transparent materials, diamond is extreme in the slowness of its light-transmission speed.

The brilliance of cut diamonds is because the light that enters a cut gem is scattered back out of the crown of the faceted gem. When light in a "slow-light" material like diamond reaches an interface between air and the gem, it either reflects back into the stone or escapes at a scattered angle. For materials that have slow light (also called high refractive index, the ratio of the speed of light in a vacuum to what it is in the material), light efficiently scatters off the lower internal parts of the stone and then exits the top. The slow speed of light in diamond enhances its efficiency in making these internal reflections off the base of the gem, its pavilion. The speed of light in diamonds varies with the wavelength or color of light. This effect is called dispersion, and blue light travels slower than red. The spectacular flashes of color seen from a faceted diamond are due to this dispersion effect (figure 7.8).

You would think that making diamonds in the lab would be impossible, given the extreme environment necessary to form them, but they have been produced, and techniques are only getting better. Because of their value and utility, there has quite naturally been a great incentive to make "synthetic diamonds" by industrial processes. At first, only tiny diamonds could be made because of the difficulty of creating the right conditions (after all, in nature the conditions necessary to create diamonds usually occur only deep beneath Earth's surface or in the craters made by large bodies impacting from space), but over time new techniques have evolved to produce larger stones, and it is now possible to produce flawless diamonds of very large size and with an

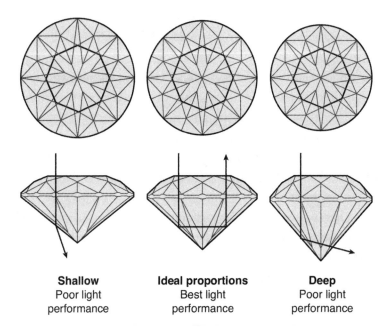

Shallow
Poor light
performance

Ideal proportions
Best light
performance

Deep
Poor light
performance

FIG 7.8. Diamond cuts to maximize brilliance.

amazing range of colors. Strongly colored diamonds, also called fancy diamonds, are exceedingly rare in nature, but this is not the case for human-made diamonds. Designer diamonds of different colors can readily be made by including tiny amounts of other atoms in their structure. And the gems produced are of very high quality. The difficulty of distinguishing natural from manufactured diamonds has caused some consternation in the diamond industry, prompting Canada, for example, to mark their diamonds with a tiny laser-engraved polar bear and a number so that they can be verified as true, natural Canadian gems.

Many synthetic diamonds are made at high pressure and high temperature in special presses. These are known as HPHT diamonds, but there are other ways to make diamonds. Very small diamonds used as polishing and grinding powders can be commercially made in explosions, and diamond coatings and thin films are made by chemical vapor deposition (CVD) processes. Instead of high pressure, CVD is done in a near vacuum in a microwave oven, where diamonds form directly from ionized gas. The conditions to make synthetic diamonds are also used to explore the nature of materials at high pressure deep inside Earth and other planets. Diamond anvils are hand-size devices that squeeze two diamonds together to produce pressures similar to that at Earth's center.

When we hear the word "diamond," many of us think of gems, Audrey Hepburn standing in front of Tiffany's, or Marilyn Monroe singing about her best friend. The gem business is the most famous part of the global diamond industry, but it represents only a part of the many ways that diamonds influence our lives. Go into almost any hardware store, and you will find abundant and relatively cheap tools that use less-than-gem-quality industrial diamonds. Diamond-based tools are becoming more and more affordable, and they are replacing others because of the superlative properties of diamond. An ordinary circular saw with an inexpensive steel blade that contains diamonds can cut granite, concrete, and ceramic tile almost like butter. The annual production of human-made industrial diamonds is billions of carats a year, and many sell for just a few dollars a carat.

Much of the utility of industrial diamonds centers on their hardness. Diamonds mark the top of the Mohs hardness scale,[4] in which hardness levels are ranked in the pecking order of who can scratch whom. Diamond is a 10, and it can scratch sapphire, next in line at 9. Quartz is a 7 and can scratch ordinary glass, which is 5.5. Gypsum is a 2 and can be scratched by a fingernail.

The great availability of inexpensive industrial diamonds has enabled modern people to do many things. The current popularity of granite countertops in kitchens is due to the affordable diamond tools that can cut large slabs of hard rock and form them into highly polished countertops with cutouts for sinks and faucets. In the past, such things were limited to royalty and the wealthiest of people, but now, due to the availability of inexpensive diamonds, these formerly luxurious stone creations can be owned by people of more modest income. Nearly all kitchen and bathroom ceramic tiles are cut with diamonds, and the list goes on. The production of highly polished concrete floors, grooved concrete highways, and airport runways all involve cutting with diamonds.

Diamonds are used in electronics and optics and to form and polish eyeglass lenses and sometimes even large telescope mirrors. Eye surgery is done with tiny scalpel blades made of diamond, because diamond knives can be sharpened to nearly atomic sharpness, and diamond knives stay sharp longer than any other kind because of their hardness. Surgical diamond knives are made from high-quality gem diamonds and are quite expensive. You can, however, go to any hardware store and get a diamond-studded file to sharpen all your steel or even ceramic knives to a supersharp edge. Some people think that electronics of the future may be based on thin diamond films, because diamond has properties that cannot be matched by the silicon-based electronic components in use at the present time.

Here we've explored why diamonds are the "superstar" form of the sixth element and seen how some of their superlative properties are unmatched by anything else that we know of. In addition to all this, they also provide a unique transport process that magically brings the deepest samples we have of our planet to the surface. Even more so than gold, diamonds have long held a special enchantment to humans. To borrow a quote from Humphrey Bogart's Sam Spade, "They are the stuff that dreams are made of." They are among the rarest things on Earth, and we wear them on our fingers, mount them in royal crowns, write songs about them, protect them in safes, and ogle them in Tiffany's. In addition, because of their unique properties, diamonds are used to drive manufacturing processes, make pursuing our hobbies and crafts easier, streamline medical procedures, and enhance our lives in countless other, often unrecognized ways.

CHAPTER 8

The Atmosphere, Climate, and Habitability

Now we turn to carbon in its softest form, as a gas, the form of carbon that most people hear about virtually every day in the context of climate change. Although only a minuscule fraction of our planet's carbon is in the atmosphere, and it is only a trace component of our air, its effects are immense. In addition to the roles it plays in our planet's climate and in long-term planetary habitability, its presence allows plants on land and phytoplankton in the sea to grow and be the base of the food chain for life on Earth.

The wondrous element carbon has benefited humans in extraordinary ways. Learning how to make fire, probably over a million years ago, clearly advanced humans beyond all other creatures on Earth. We could use it to keep warm, cook our food, and make farmlands out of forests. Oxidation of carbon to provide energy ultimately evolved to the large-scale harvest of our planet's vast subterranean reservoirs of fossil fuels. This "gift of nature" powered the Industrial Revolution and provided the energy to make the modern world we live in. Perhaps humans would never have advanced beyond other animals without the process of burning carbon compounds into carbon dioxide. The abundant energy also allowed the human population to increase to 8 billion people. The combination of unfettered population growth and unprecedented burning of carbon has led to the realization that our playing with fire has a drastic consequence.[8.1] The rise of carbon dioxide is leading to a level of global warming that seriously threatens our modern, overpopulated world. Over the last century, the Earth's average surface temperature has increased by about 0.1°C each decade, and the rate has accelerated over the past few decades. Much of the increase is commonly associated with the rise of carbon dioxide and its role in the greenhouse effect, a process that formerly seemed arcane but is now known to be vitally important and is constantly in the news.

Recognition of the importance of the greenhouse effect was slow to develop. The greenhouse effect was first noted as long ago as the 1820s, although it wasn't referred to by that name. Joseph Fourier was primarily a mathematician and physicist and is most famous for showing that any mathematical function can be matched by adding sine waves of different frequencies, amplitudes, and phases. Late in life he took an interest in atmospheric heating, and he showed that Earth's atmosphere is warmer than it should be. He realized that some process must keep it warm instead of being so cold that it couldn't support life.

Fourier thought of a process that traps heat by causing a surface to warm and emit "invisible rays." When the wavelengths of these rays were discovered, they were found to be longer than those of visible light, and they were called infrared radiation because they were beyond the red end of the visible spectrum.

Before 1900, Swedish scientist Arvid Högbom noticed that CO_2 emitted from factories raised the CO_2 level in the atmosphere. A colleague of Högbom, Svante Arrhenius, previously mentioned for his suggestions on panspermia as well as life on Venus, thought that the ocean's intake would balance the CO_2 emissions and prevent atmospheric buildup. Carbon dioxide is absorbed in ocean surfaces, and it was thought for some time that this would stabilize the temperature at a comfortable level. This is true, but not on the right time scale. The average depth of the oceans is nearly 4 kilometers (almost 2.5 miles), and it takes a long time for something to mix from the ocean's surface to its bottom.

A "green" house is one having south-facing windows (in the Northern Hemisphere) so that sunlight coming in heats the interior; the resulting infrared radiation is trapped. In a perfect, ideal greenhouse, the light enters but the emitted infrared does not get out. The greenhouse effect could just as well be called the "auto-glass effect," because it is a similar phenomenon that causes a car parked in a sunny spot to be warmer inside than the outside temperature. This is why you don't leave your children, dogs, or chocolate candy bars in a closed car on a sunny day.

The "greenhouse effect" isn't really a very exact metaphor. In a garden greenhouse, the outgoing radiation is stopped by panes of glass; in a global "greenhouse," the atmosphere absorbs infrared (IR) radiation throughout its height. It's not a house but an atmosphere. Truth be told, a major reason that garden greenhouses are warm is that they block the wind. Heat is trapped inside; it doesn't blow away and is not diluted by cold air blowing in. The enclosure inhibits convection, which is the dominant form of vertical heat transfer in the lower atmosphere.

The atmospheric greenhouse effect works in the following general way: sunlight entering the atmosphere reaches the ground and heats it. The ground takes some of this energy and emits it skyward as infrared heat radiation. This emitted radiation bounces around in all directions as it is absorbed and reemitted by air molecules many times over, in a so-called "random walk." The heat eventually escapes, but on the way all this bouncing around transfers energy to the air. This is how the greenhouse effect causes a planet's surface to be warmer than it would be without it.[8.7, 8.8]

Carbon dioxide is not the only culprit in global warming. Water vapor, methane, ozone, and various chlorofluorocarbons are also involved and absorb infrared. These gases are actually more efficient in absorbing heat than CO_2, but, with the exception of water vapor, they are less abundant. Water vapor is the dominant cause of the Earth's overall greenhouse heating, but carbon dioxide is largely responsible for the present increase in atmospheric heating because its abundance is continuously and rapidly growing. Changes in the abundance of this carbon-bearing molecule are a major influence on climate stability and instability, and this will continue for years to come.

On the other hand, greenhouse warming and the role of CO_2 variations have also played a positive central role in promoting Earth's habitability over very long periods of time. One quite remarkable aspect of its history is that Earth has been at least moderately stable for much of its lifetime, in great contrast to our neighbors, Mars and Venus. Mars seems to have been a warmer, wetter place billions of years ago, but it has been a frozen desert for the last 3 billion years, and Venus is a hellish place that seems to have lost all of the surface features it had for the first 3.5 billion years of its existence. (Venus likely had an ocean like ours but then lost it to space.) Other than a few rare Snowball Earth episodes, the Earth's complex stabilization systems have kept the ship fairly steady. This stability is highly advantageous for the survival of higher organisms, because it minimizes abrupt environmental changes that lead to major extinction events. In fact, the Earth's environmental stability is extraordinary, considering that many other things have changed. The Sun has continued to get brighter by about 10 percent every billion years. The composition of our atmosphere has profoundly changed over geological time scales. Even though carbon dioxide is rapidly increasing at present, over Earth's history it generally declined. Free oxygen is abundant in our modern atmosphere, but it was essentially nonexistent before 2.5 billion years ago. Our preoxygen atmosphere may have contained significant amounts of methane, a strong greenhouse gas. Among other changes are the ratio of land to ocean area and the spatial distribution of land. During Earth's greatest known extinction, the Permian-Triassic extinction event, or the "Great Dying," 250 million years ago, most land area was in the same region. The early Earth may have been a "water world" with just a few islands, and the continents we have today may have slowly built up over time.

Unlike Mars and Venus, Earth's lithosphere, its "rigid" outer layer, is not static. The continents and ocean floor are parts of plates that move, a phenomenon noted by scientists since the sixteenth century. Alfred Wegener made a methodical study of continental drift, as it was known, and published a scientific paper in 1912, in which he cited the evidence for continent movement.

Our planet's plate movements are ultimately driven by the internal heat left over from Earth's formation and produced by radioactive decay. Some plate boundaries are spreading centers, or divergent boundaries, where new igneous rock is formed from cooling magma and moves outward at rates of centimeters a year. As the material cools, it becomes denser, and when it ultimately reaches a convergent boundary with a continental plate, it is diverted downward in a process called subduction, descending to depths where it is transformed by heat and pressure. Unique to Earth, this remarkable process occurs on a hundred-million-year time scale and forms the famous conveyor belt through which oceanic crustal material is formed and later destroyed. Because of this recycling process, the oldest rocks on the ocean floor are less than 5 percent of the age of our planet.

An important long-term geologic cycle called the carbonate-silicate cycle, mentioned in chapter 4 and elsewhere, has acted as a thermostat to help regulate surface temperature on geologic time scales, and the global recycling of matter by plate tectonics is a key feature that enables this thermostat to work its wonders. Earth's carbonate-silicate cycle starts with rain, which contains carbonic acid (H_2CO_3) from CO_2 in the atmosphere. The weak acid dissolves calcium from rocks, and some calcium-bearing ions reach the ocean, where they are used to make shells and skeletons that end up on the sea floor, forming carbonate deposits such as limestone. This process is a "sink" for atmospheric carbon because it removes CO_2. The subduction process forces oceanic crust below continental margins where those carbonates thermally decompose and release CO_2 to the atmosphere via volcanic activity. Without the recycling of plate tectonics, the chemical weathering process would remove our atmospheric CO_2 to the point where plants could not grow, as the much-maligned CO_2 in our atmosphere is actually Earth's fundamental "food of life." As long as the plates move, however, there is a steady source of CO_2 going back into the atmosphere. (It is possible that planets lacking plate tectonics could have volcanic processes that also might drive a carbonate-silicate cycle, but it would be unlike what happens here.)

This process acts like a thermostat because the weathering process that removes CO_2 to form new carbonates is temperature dependent. When the planet cools, the rate of weathering diminishes and allows CO_2 to accumulate in the atmosphere and promote warming. When the planet is hot, the weathering rate increases and removes more CO_2 from the air to form carbonate rocks like limestone and promote cooling. This regulatory feedback of carbon in the atmosphere is believed to be a major factor that has kept our planet's climate reasonably stable for billions of years. Unfortunately, this wonderful natural thermostat is too slow to counteract the buildup from the burning of fossil fuels.

Earth can be considered to have two major carbon cycles: the slow carbonate-silicate cycle we just described and a faster one, often called the short-term cycle, that is driven by biological processes. The short-term cycle is synchronized with the seasons as plants grow and die, taking in carbon and releasing oxygen in a yearly cycle. This cycle is seen as a small annual oscillation of atmospheric CO_2.[1] Sunlight is crucial, providing energy for the growth. In chemical terms, this is one sequence of reactions:

$$6CO_2 + 6H_2O + \text{energy} = C_6H_{12}O_6 + 6O_2$$

And then:

$$C_6H_{12}O_6 + 6O_2 = 6CO_2 + 6H_2O + \text{energy}$$

The molecule $C_6H_{12}O_6$ is a sugar, and the energy to produce it comes from sunlight. The short cycle is easy to observe, especially in midlatitudes. The annual change of the seasons, with the surge of plant growth alternating between the Northern and Southern Hemispheres, is due to the short-term cycle.

For good or bad, there has long been a strong link between life and the evolution of our atmosphere.[8.2] There is rapid change occurring in our atmosphere right now, driven largely by carbon dioxide buildup from a single species—us. Paul Crutzen suggested that we consider our planet to be in a new human-dominated geologic epoch that he calls the Anthropocene, which follows the Holocene, the twelve-thousand-year epoch that began when the last glacial period ended.[8.3] The cause of the Anthropocene is the unfettered growth of the human population and the large-scale utilization and modification of Earth's resources. In a 2002 *Nature* paper, Crutzen stated, "Unless there is a global catastrophe—a meteorite impact, a world war or pandemic—mankind will remain a major environmental force for many millennia."[8.3] A much more dour assessment of our current predicament was given by Sir Martin Rees in his book *Our Final Hour: A Scientist's Warning: How Terror, Error, and Environmental Disaster Threaten Humankind's Future in This Century—On Earth and Beyond* (Basic Books, 2004), which describes numerous ways that humans might cause their own extinction within this century.

The recent growth of atmospheric CO_2 is spectacularly shown by measurements made from the Mauna Loa Observatory at the summit of Mauna Loa on the Big Island of Hawaii.[8.4] Sixty years of continuous measurements show a steady increase of about 30 percent over 1960 values. Superimposed on the long-term trend is an annual variation of about 2 percent from peak to peak due to the seasonal growth of Northern Hemisphere land plants, the previously mentioned short-term cycle. The plot is called the Keeling Curve after the Scripps Institution scientist who initiated the ongoing measurements in 1958 (figure 8.1). Most scientists, especially climatologists, believe that this rise, clearly shown in the Keeling Curve, has led to a self-inflicted

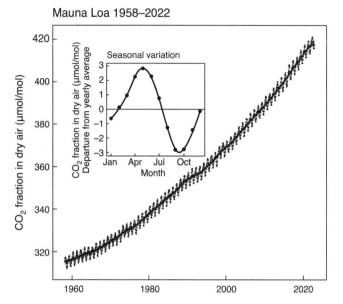

FIG 8.1. The famous Keeling Curve showing the rise of atmospheric CO_2 over time from the summit of Mauna Loa in Hawaii. The insert shows the six-parts-per-million seasonal variation, due to the increased consumption of CO_2 for photosynthesis in the spring and summer "growing seasons." *Credit:* Data from Dr. Pieter Tans, NOAA/ESRL, and Dr. Ralph Keeling, Scripps Institution of Oceanography.

climate crisis or even catastrophe. Already we are seeing melting glaciers, record temperatures, and more severe storms. This trend will affect crops, forests, and many other things influenced by climate change. Major crop growing zones will change, including shifts to more northern or southern regions, and many of the world's deserts will get larger. The societal and economic impacts of these changes could be disastrous in some regions of the globe. Climate and sea level have always fluctuated over Earth's history, but the abruptness of the current global warming appears to be unprecedented over the past few millions of years.

Almost everyone has an opinion about global warming. Most scientists attribute the current increase to carbon dioxide with contributions from other greenhouse gases, such as methane, nitrous oxide, and ozone. There are a few skeptics, who point to fluctuations in the long history of Earth. But by now the evidence is clear: the warming over the last century is the result of human activities that put more carbon into the atmosphere than is being removed. We are causing this change, and we and several generations that will follow us will have to live with the consequences of our rapid burning of fossil fuel.

Though the current warming period is ushering in big changes on our planet, Earth has endured big changes before. Past dramatic temperature changes have included warm periods and, of course, glacial periods. Just twelve thousand years ago, Seattle, Chicago, and New York were covered with a kilometer of glacial ice. Humans survived this glacial period, and for the last twelve thousand years, since the dawn of civilization, we have been enjoying a comparatively warm "interglacial" period. Over the last million years, Earth has seen nearly a dozen cold periods where ice formed well beyond polar regions. During this time, glacial periods were actually the norm, and each lasted roughly one hundred thousand years. Averaged over time, only a small fraction of the last million years had mid-latitude temperatures as warm as we have experienced during recorded history.

Global mean temperatures vary over recent and geological time periods. The dinosaurs lived in the Mesozoic Era between 66 and 245 million years ago, when global temperatures were much warmer than present. Note the periods with and without polar ice caps in figure 8.2.

We are surely in trying times with our human-produced sharp increase of global warming, but viewed in broad perspective,

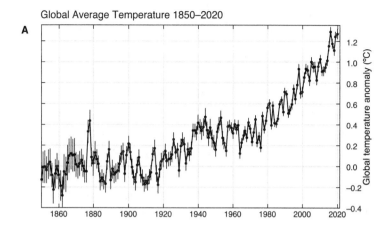

Global Average Temperature 1850–2020

FIG 8.2A. Global average temperature, 1850–2020. Land data prepared by Berkeley Earth and combined with ocean data adapted from the UK Hadley Centre. Global temperature anomalies are relative to the 1850–1900 average; vertical lines indicate 95 percent confidence intervals. Credit: "Global Temperature Report for 2020" by Robert Rohde. © Berkeley Earth.

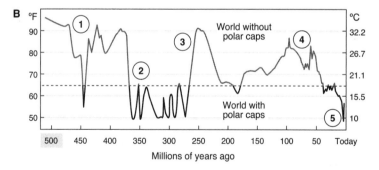

FIG 8.2B. Global mean temperatures vary over recent and geological time periods. The dinosaurs lived in the Mesozoic Era between 66 and 245 million years ago when global temperatures were much warmer than the present. Note the periods with and without polar ice caps. From "A 500-Million-Year Survey of Earth's Climate Reveals Dire Warning for Humanity" by Paul Voosen. Credit: Smithsonian Institution National Museum of Natural History, adapted by N. Desai/Science.

some level of change on our planet is the norm. Over long periods of time, Earth fluctuates between warmer "hothouse" states (often called "greenhouse" states)[8.5] and cooler "icehouse" states. Generally, the hot/cold states correlate with periods of higher or lower abundances of atmospheric carbon dioxide and other factors. In its greenhouse state, Earth is too warm for glaciers to form on continents or at the poles, as shown in figure 8.2. Earth has been in its hot greenhouse state for about 85 percent of its overall history.

When not in a hot greenhouse state, our planet is considered to be in an icehouse. There have been five recognized icehouse periods on Earth including the Late Cenozoic since the modern Antarctic ice sheets began to form nearly 34 million years ago. All of human history has occurred during an icehouse period in which both poles have been covered with ice and global climate has been moderate. Icehouse periods can feature "glacial" and "interglacial" episodes during which ice sheets can grow or retreat. The rise of civilization has occurred in an interglacial period, but for over 80 percent of the past few million years our planet has been in cooler glacial periods. The switch between glacial and interglacial happens on time scales of roughly 0.1 million years and is generally thought to be influenced by astronomical factors such as small changes in Earth's orbit around the Sun and the tilt of its spin axis.

There have been more spectacular temperature fluctuations a few times in Earth's history. A little more than 500 million years ago, and also about 2.5 billion years ago, our planet's climate is believed to have become quite unstable, with ice forming in the equatorial regions. In these Snowball Earth episodes, the oceans largely froze over, thwarting the removal of atmospheric CO_2 by its normal path of dissolving in sea water. Some have estimated that CO_2 released by volcanic activity might have

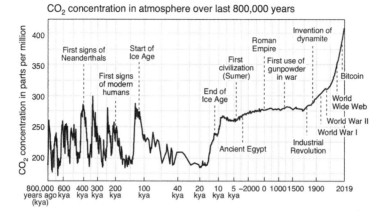

FIG 8.3. CO_2 change over the past eight hundred thousand years. *Credit:* NOAA and NASA.

built up in the atmosphere to levels of 100,000 ppm by the end of a Snowball episode. This is more than a hundred times higher than the current level of 420 ppm. Current theory posits that when such an episode ended, great amounts of accumulated carbon dioxide caused temperatures to shoot to hellish values before it was removed by chemical processes and returned to the ocean and subterranean reservoirs. The extremely high temperatures accelerated the chemical weathering processes that remove carbon dioxide from the atmosphere. The ocean-chemical changes during deglaciation periods sometimes resulted in the formation of thick limestone deposits, called cap carbonates. These were deposited on top of geological formations made by glaciers or material dropped by ice.

Looking at CO_2 over the last million years provides an interesting perspective on natural cycles as well as the effects due to human activity (figure 8.3). Twenty thousand years ago, at last glacial maximum, carbon dioxide abundance was about 200 ppm, and 120,000 years ago, at the peak of the last interglacial warm period, it was 280 ppm. Averaged over the past 10,000 years, carbon dioxide abundance has been around 280 ppm, but a huge spike began with the Industrial Revolution, and CO_2 is presently the highest that it has been over the last 800,000 years. We have slowly gotten ourselves into a real pickle, and it will be an enormous challenge for us to manage and attempt to minimize the myriad changes that the rise in CO_2 will cause.[8.6]

The rapid rise of CO_2 causes serious problems for humans because of the ways it changes the climate, the acidity of the oceans, and sea levels. Our food supply, our fresh water supply, our weather, and many other issues are influenced by rising CO_2. The carbon dioxide level will continue to rise and, as mentioned, has already increased by more than 30 percent over the last half century. At the current projected rate of increase, we will reach about 500 ppm fifty years from now, enough to raise sea levels about one meter by 2100 and a few meters by 2300. The main concern regarding the rapid increase is the strain it puts on Earth's complex mix of systems that involve air, land, water, and organisms. However, the direct effects of inhaling more CO_2 will not be a problem, as least for the next few generations. Carbon dioxide levels inside present-day homes can exceed 1,000 ppm and are about 5 percent (50,000 ppm) in the air that we exhale. The CO_2 level inside COVID-19 masks often exceeds 5,000 ppm, but this has not been considered to be a health hazard (although levels above 5,000 ppm for over eight hours are considered to be a problem).

Even if all fossil fuel emissions stopped right now, sea levels would continue to rise due to the inertia of the earthly processes involved. Some estimates predict even larger rates of increases as ice melts and water warms and expands. Other estimates suggest a slower rate of increase. Rising seas will cause flooding in coastal areas, such as Bangladesh, Kolkata, Mumbai, and Dhaka, and the Netherlands will have to raise their dikes even higher. Twenty thousand years ago, during the last glaciation, the sea level was more than 100 meters lower. Our hunter-gatherer ancestors must have lost some living areas as the oceans rose to current levels, but they did not have to deal with the extensive infrastructure that our modern world uses to support its human population of 8 billion.

The coming sea level projections are based on estimates of gradual warming, but some models suggest that the situation may be more chaotic. If entire Antarctic ice shelves change too much, nonlinear processes may take over, perhaps raising the global temperature dramatically and accelerating sea level rise.

James Hansen, now an adjunct professor at Columbia University, stirred up considerable controversy with his dire warnings of imminent disaster if nothing was done to slow the warming—or at least to stop adding to it. He testified before a 1988 Senate committee on energy and natural resources that there was a high degree of confidence that the accumulation of human-produced greenhouse gases was causing global warming. Many consider that his testimony was a watershed moment in drawing international attention to global warming. When Hansen was a NASA employee, he was subject to censorship by government intervention, but the scientific community put enough pressure on the government to change that. In 2007, Al Gore and the UN's Intergovernmental Panel on Climate Change won the Nobel Peace Prize for their work on climate change, a sign of public recognition of global change.

The greenhouse effect actually provides a globally averaged long-term stable background. In contrast, the dangerous anthropogenic greenhouse gases do not condense but just continue to build up in the atmosphere on human time scales (figure 8.4). We are far enough from the Sun that we would

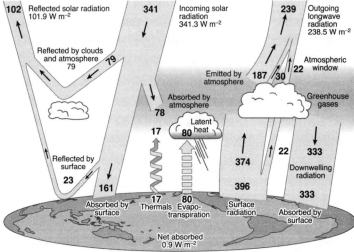

FIG 8.4. Greenhouse heating. The power (watts per square meter) that Earth receives from the Sun is matched by the power that is both reflected and radiated back into space, but the atmospheric temperature is influenced by greenhouse gas that absorbs infrared light. *Credit:* From "Earth's Global Energy Budget," by Kevin E. Trenberth, John T. Fasullo, and Jeffrey Kiehl (*Bulletin of the American Meteorological Society* 90, no. 3) © 2009 American Meteorological Society.

always be below freezing if we didn't have water vapor and the rest of the greenhouse gang in our air. Water vapor is the major source of the greenhouse effect that warms Earth by about 30°C/86°F, but it does not play a primary role in the current atmospheric temperature increase caused by human activity. With anthropogenic heating, water vapor does function as a feedback factor that amplifies the effects of increases of noncondensable greenhouse gases like CO_2, CH_4, and N_2O. As the planet warms even by a small amount, the atmosphere can retain a higher abundance of water vapor.

Besides the dominant nitrogen and oxygen, the next most common molecules and atoms in our air are H_2O, argon, and CO_2, in that order. One percent of our air is argon that is produced by radioactive decay of potassium. It is twenty times more abundant than CO_2, but as an inert gas, it has no significant effect on climate, global warming, or us.

Most of the Earth's near-surface carbon is tied up in carbonate rocks like limestone and dolomite, but it can only interact with the oceans and the atmosphere on geological time scales, over millions of years. Carbon dioxide that can interact with us over shorter times, thousands of years or less, is held in the oceans. About 93 percent of the total carbon dioxide is in the oceans. Carbon dioxide rapidly dissolves in water, where it reacts with water to form carbonic acid (H_2CO_3), which rapidly ionizes to form bicarbonate (HCO_3^-) and carbonate (CO_3^2) and eventually sinks to the sea floor.

The cycle of carbon also includes biomass, the amount held by plants and animals, which is only about 0.04 percent of the carbon near Earth's surface. You would think that's too small to have any effect on atmospheric temperatures. In contrast, CO_2 above ground is very important because it affects the surface temperature, where we live.

Almost everybody knows that injecting carbon dioxide and methane into the atmosphere is bad and causes global temperatures to rise. A global focus on decarbonization and ending our historic dependence on energy and materials derived from geological carbon materials has given rise to concern about "carbon footprints" and "carbon credits" exchanged between carbon-emitting companies to help reduce human energy use. International politics is affected as well, with most nations trying to reduce the carbon emissions from industrial sources and general population sources. In some countries, forests are burned to make pasture for crops and ranches. There you lose twice: carbon is injected into the atmosphere, and at the same time there is less forest to convert CO_2 into free oxygen by photosynthesis.

Industrialized nations contribute most of the greenhouse gases. China, the United States, and the European Union emit more than half of the world carbon dioxide from fossil fuels. The Paris Agreement of 2016 is an accord between 196 parties at the UN Climate Change Conference (COP21) to stabilize carbon emissions and limit average temperature increase, ideally to about 1.5°C/2.7°F. The US backed out of the agreement in 2020 but rejoined 107 days later after a change in US presidential administrations. The main goal is to limit greenhouse gas emissions from human activity between 2050 and 2100 to the levels that trees, soil, and oceans can absorb naturally. This balance is known as "net zero," and many ideas have been put forward to help us evolve to net zero emissions.[8.9]

As a mental exercise, we can imagine an Earth with an extreme greenhouse change: zero greenhouse gas in the atmosphere. Then it's easy to calculate that planet's surface temperature, because sunlight would just come through to the surface without being dimmed. In fact, we have an example: the Moon. We can see how much the sunlight heats Moon's surface with no atmosphere and zero greenhouse warming. Such a calculation leads to an average surface temperature below freezing! The actual temperatures on the lunar surface range from well below freezing to above the boiling point of water under 1 atmosphere of pressure. It gets very hot in the noonday Sun but extremely cold at night because the ground radiates its energy into the blackness of space without any greenhouse gas to stop it. The Moon is lifeless, and it effectively has no atmosphere, no liquid water, and no natural habitability for any known organism.

Earth's atmospheric carbon is unique among all the chemical elements in the attention that it gets from concerned peoples and governments across the globe. The human population is placing unprecedented focus on the sixth element and spending trillions of dollars to reduce its effects on climate, maintain our planet's remarkable habitability, and keep Earth as the solar system's singular "garden planet." Earth is a complex place, and understanding its workings and predicting its future are daunting challenges. An important factor in enabling reliable predictions is detailed monitoring of Earth and its multitude of interacting processes. Examples of global monitoring programs are the NASA Orbiting Carbon Observatories (OCO-2 and OCO-3); the international program Deep Carbon Observatory (DCO); and the European Space Agency Copernicus program that will launch a multibillion-dollar Earth monitoring project, called Copernicus Anthropogenic Carbon Dioxide Monitoring (CO2M), that will involve several orbiting spacecraft.

OCO-2 is in a polar orbit, going north to south, as the Earth turns under it. OCO-3 is a suite of instruments that were spares for OCO-2 and are mounted on the International Space Station. The amount of CO_2, methane, and oxygen is measured and mapped in both time and location. For instance, the oceanic warming occurring every few years, the so-called El Niño oscillation, or the effects of large forest fires, can help us understand the global environment.

The Deep Carbon Observatory (DCO) involves atmospheric scientists, geologists, geochemists, mineralogists, oceanographers, and more. Every scientific field related to carbon has someone represented on the team. As described by its executive leader, Robert Hazen, DCO is "dedicated to understanding the quantities, the forms, the origins and the movements of carbon from crust to core in planet Earth."[8.10]

One goal of the DCO is to find all carbon-bearing minerals. (On Earth, there are no free carbon atoms; as discussed, free carbon atoms exist only in the interstellar medium. Here, all carbon in solid form is combined with itself or with other elements to form minerals or compounds.) At present, there are over 403 carbon-bearing minerals, and more than 100 are predicted but not found yet. In 2015, Hazen organized a citizen science project called the Carbon Mineral Challenge to encourage the identification of new carbon minerals. This led to the discovery of over thirty new minerals, including Edscottite, an iron carbon mineral in meteorites, named after Ed Scott, a famed meteorite researcher at the University of Hawaii.

However bad things may become, it seems quite improbable that, as frequently implied, the current global warming event will lead to the end of the world or even simply extinguishing life on our planet. Planets are quite difficult to destroy, and extinguishing all life has not occurred over billions of years of habitation. Plants and animals are somewhat fragile, but microbial life is robust. Once a planet is infected by microbial life, it takes extreme processes to globally sterilize it. Life can probably survive any level of damage that humans could ever do to our planet. Life's end and even Earth's end will eventually come, but only in the distant future, when well-understood processes inevitably cause the Sun to become brighter and larger.

Life has endured climate changes in the past, although not without several very serious mass extinction events.[8.11] The dinosaurs lived for almost 200 million years in warmer and more carbon-rich air than we do, but they had evolved to thrive in the environment of the Mesozoic Era. They became extinct because of environmental change initiated by the chance impact of a 10-km rock from space. This event caused the extinction of 75 percent of animal species on Earth and all four-limbed animals larger than 25 kg. Earth has seen even worse mass extinction events, but life has survived and carried on. Long before the first dinosaurs, 500 million years ago, the atmosphere contained over ten times as much CO_2 as it does now. Life also managed to survive the extreme temperature and CO_2 changes that are believed to have occurred during Snowball Earth episodes.

CHAPTER 9

Carbon Out There

Our Galaxy is a giant carbon-producing factory. Star formation can occur in bursts in different regions, but the net result is making more carbon and other heavy elements. It is still ongoing, but the peak of star formation occurred early in the history of the Galaxy. From the perspective of a book on carbon, a major point of interest is the production of carbon in stellar cores and its recycling between stars and the interstellar medium.

Galaxies are collections of thousands to over a trillion stars, bound together by gravity. As discussed earlier, Edwin Hubble was the first astronomer to prove that some of the faint clouds seen in telescopes are other galaxies, outside the Milky Way. He made a diagram based on their shapes. We live in a spiral barred galaxy of intermediate size, in a spiral arm. It was difficult to deduce that, because we are inside it. It is hard to see the forest through the trees.

The most visible galaxy that can be seen with the naked eye from the Northern Hemisphere is Andromeda, or M31 (figure 9.1). It is the most distant object that you can see just with your eyes, and the light you see took 2.5 million years to travel this far. Swiss American astronomer Walter Baade was able observe

FIG 9.1. Andromeda Galaxy, a companion of the Milky Way. It is easily seen with the naked eye in dark skies, and it is the most distant object that can be seen with the naked eye. *Credit:* David (Deddy) Dayag via Wikimedia Commons (CC BY-SA 4.0).

M31 at high resolution,[1] and he realized the stars within the galaxy could be classified into distinct groups—stars in spiral arms were named "Population I," and stars in the bulge and halo outside the disk were named "Population II."

Our Galaxy, the Milky Way, began to form early in the history of the Universe. Initially, it was a gravitationally bound mass of gas and stars that grew as it accumulated much of its new mass by mergers with other galaxies, large and small. The Milky Way is still accumulating small nearby galaxies, and a few billion years from now, we will merge with Andromeda, which is coming our way.

As the Milky Way flattened to become more disk-like, some stars were left behind, forming a spherical "halo." Globular clusters containing thousands of stars formed in the halo. The first globular cluster where individual stars could be resolved is M4. It is visible with the naked eye near the star Antares. Globular cluster stars formed at the Galaxy's birth, or accreted from external galaxies, and are uniformly old. These are Population II stars, and they have much less carbon, iron, and other heavy elements than Population I stars like our Sun—some a thousand times less. They are poor in heavy elements because they formed before element formation in stars and recycling back into the interstellar medium had built up an accumulation of such elements in the Galaxy. Another group of stars are the "high-velocity" stars, whizzing by our solar system at speeds up to 100 to 200 km/sec. They come from every direction in the halo, not the galactic plane; now they are recognized as Population II stars.

There is a group of stars even older than the Population II, naturally labeled Population III, that have no elements heavier than hydrogen and helium, and thus no carbon yet. Until mid-2023, no pure Population III star had been found, only stars with *almost* no carbon. These stars are labeled "metal-poor,"[9.1] and only a few had been observed in the outer Milky Way. Finding them is an important goal of the large NASA James Webb Space Telescope (JWST) put into operation in 2022. This new space telescope is revolutionizing our ability to explore the most distant and earliest formed objects.[9.2, 9.3, 9.4, 9.5]

In mid-2023, three teams of astronomers, using different techniques, reported what appears to be the first evidence of Population III stars. Two detections were made by JWST, which observed galaxies with very high redshifts, meaning they formed early in the universe, and found stars made of only hydrogen and helium. The third observation, by ground-based radio telescopes using a very indirect method, supported the same conclusion.

Due to the expansion of the universe, explained in chapter 1, the distance to a galaxy can be derived from its spectrum. Its spectral lines are shifted toward the red to longer wavelengths than their rest wavelengths. Stars in the Milky Way have low redshifts, because they are not moving away very fast. Faraway galaxies have bigger redshifts—much bigger redshifts in some cases. The value of redshift is called z where z is defined by:

$$z = \frac{\text{observed wavelength } - \text{rest wavelength}}{\text{rest wavelength}}$$

The highest z observed so far is $z = 13.27$, which means the galaxy is so far away that the light we see from it left it only 300 million years after the Big Bang.

These first stars were massive, up to a few hundred times the Sun's mass. As explained in chapter 1, massive stars have short lifetimes, so these stars quickly exploded as supernovas, making new elements in the process. When the universe got clumpy and cold enough, molecules could form. We don't know yet exactly when that occurred, but astronomers, using the JWST, detected PAHs, described in chapter 2, in a galaxy formed 1.5 billion years after the Big Bang. Presumably, more complex molecules, described in chapter 5, were created then as well. The Webb telescope was designed to study the earliest history of star formation, and it is expected that continued studies will provide great insight into the first formation of carbon and other heavy elements.

In addition to stars, galaxies can also contain black holes, very massive ones. After World War II, radio telescopes were developed to explore the sky at radio wavelengths. Astronomers at Cambridge University in England made several all-sky surveys and found several unknown sources of radio emissions. At first, radio sources were difficult to link to optically observable

FIG 9.2. Hubble's Wide Field and Planetary Camera 2 (WFPC2) took this picture—likely the best of ancient and brilliant quasar 3C 273—which resides in a giant elliptical galaxy in the constellation of Virgo. It was the first quasar ever to be identified and was discovered by astronomer Allan Sandage. *Credit:* ESA/Hubble and NASA.

sources because the spatial resolution of radio telescopes was too limited. In the early 1950s, a few mysterious sources were found, and they were called QSOs (Quasi Stellar Objects). They were also called quasars for "quasi-stellar radio source" (figure 9.2). We will use both names. Their optical spectra were mysterious, not coming from any elements that scientists knew.

One of the pioneers of quasar science was Allan Sandage. While Sandage was a graduate student at Caltech in the early 1950s, Edward Hubble hired him as an assistant to help with observations. In those days, there were no digital imagers like those used in modern phones, cameras, and telescopes. Astronomers used photographic glass plates, which had to be developed in a darkroom. That was one reason Sandage was hired. On the way, he learned to observe faint stars and galaxies.

In 1963, Sandage (figure 9.3, top) was the first astronomer to realize that the radio source 3C 48 was associated with what appeared to be a star.[2] It had mysterious emission lines, unlike the ordinary absorption lines a normal star would have. Usually spectral lines in stars are dark where specific wavelengths were absorbed by gas. Emission lines are bright and are usually caused by hotter than normal gas. Sandage had no idea what the source could be. Three years later, another QSO, 3C 273, was identified as a starlike object, and it had the same pattern of emission lines as 3C 48. But this time, astronomer Maarten Schmidt (figure 9.3, bottom) realized that the spectrum was hydrogen but redshifted to longer wavelengths. 3C 273 was moving from us at 47,000 km/29,204 miles per second, or 15 percent of the speed of light. No other known source could move like that. Then 3C 48, with its own hydrogen spectrum, was found to be moving away at 92,000 km/57,166 miles per second, or 37 percent of the speed of light. Quasar spectra are always shifted toward longer wavelengths.

A spectrum of a QSO shows emission lines, implying that the gas forming them is hotter than normal. Carbon is highly ionized to a level called C IV, having lost three electrons out of its total of six. That only occurs when the gas forming the lines is much hotter than the "surface" layers of the Sun. Other elements are similarly ionized; examples include O VI (oxygen), Si IV (silicon), Fe II (iron), and Mg II (magnesium).

Along the way, light coming to us from a quasar passes through intergalactic clouds, forming absorption lines. These features always have smaller redshifts than the QSO, which tells us that these clouds are not as far away as the QSO. Again, carbon

FIG 9.3. Alan Sandage and Maarten Schmidt. *Credit:* (*top*) Carnegie Institution of Washington; (*bottom*) *Time* Magazine cover, March 11, 1966.

is heavily ionized, going to the C IV state. These clouds are heated by supernovae.

Many astronomers were initially dubious that any source could move so fast, but in time most scientists agreed with that interpretation: the QSOs were mysterious objects moving away at very high speed. The more distant that an object is, the faster it moves away from us, following the famous Hubble law. Their speed is due to the expansion of the Universe, implying they are extremely distant objects. They are very powerful, with intrinsic brightness many thousands of times that of stars.

With care and giant telescopes, astronomers learned that only about 10 percent of QSOs are strong radio sources. Now a catalog of QSOs contains more than five hundred thousand entries—they are everywhere in the sky! The record redshift, observed in 2021, is 7.642, evidence that galaxies were already formed just 670 million years after the Big Bang.

The telescopes also found that quasars are embedded in the cores of faint galaxies. The quasar is brighter, a lot brighter, than the surrounding stars and nebulae, so the first thing an astronomer sees is the quasar, not the stuff around it.

Let's summarize: the QSOs are powerful sources of radio and visible emission; they live in galaxies; they are tiny; and their life began near the Big Bang, when the first galaxies formed. To scientists, the only entity that can act like that is a black hole. Indeed, observations of nearer massive galaxies, including our own, confirm the interpretation that QSOs are black holes inside galaxies.[3] Their masses are huge, a million or more times more massive than our Sun. Most astronomers call them "supermassive" black holes, as opposed to stellar mass black holes described in chapter 1. The observed emission of QSOs at the centers of galaxies is not from the black hole but from material falling into it. Once closer than a distance called

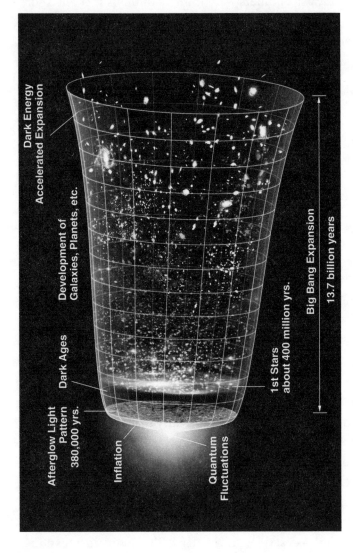

FIG 9.4. Big Bang and expansion of the Universe. *Credit:* NASA/WMAP Science Team.

the "event horizon," nothing, not even light, can escape the great gravitational pull of a black hole. Novels and movies describe how black holes gobble up everything that gets sucked in, carbon atoms, dust, planets, stars, and interstellar gas. Black holes are an amazing way that carbon atoms can be totally destroyed and effectively removed from the cosmos.

Carbon: From the Beginning to Infinity

Now we are in position to appreciate the full evolution of the Universe and the sixth element. When the Universe was young, just after the Big Bang, there was only hydrogen and helium, essentially—no carbon, no atoms heavier than lithium. As time progressed, the Universe cooled, became less dense, and a sequence of changes occurred. Early on, there was a period when the Universe was so hot that many atoms were ionized. Because atoms could absorb light and lose electrons, the Universe was opaque to visible light, and matter and light strongly interacted; light and matter were coupled to each other. After a few hundred thousand years, when the temperature dropped to about 3,000°C, the Universe finally became transparent to light. Light and matter rarely interacted after this time, the epoch of decoupling. The light from this time has now redshifted to microwave radio wavelengths, first detected in 1965 as a cosmic microwave background. This radio noise filling the sky provided the first smoking-gun evidence that the Big Bang had occurred.

At the time it became transparent, the Universe was filled with red light, but after continued expansion and cooling, the light redshifted to wavelengths too long to be visible to the human eye. The Universe entered a "dark age," where there was no leftover visible light from the Big Bang and no new sources generating light. After a period of at least a few hundred million years, conditions were finally right, and matter started to gravitationally collapse and form stars and galaxies. Stars lit up the Universe, and nuclear reactions in their hot cores created new elements, including carbon. Some of the elements recycled back into space to form new generations of stars and their planets. Over time, elements such as carbon, oxygen, silicon, magnesium, and iron became abundant enough to form Earth-like planets suitable for the origin or at least survival of organisms such as those that have existed on Earth for billions of years.

The dark age ended over 10 billion years ago, and since then, the Universe has continually been making stars, planets, and carbon, as well as other elements. This heady process goes on and on, but it can't last. Stars have lifetimes limited by their available nuclear fuel. The rate of formation of new stars to replace old ones diminishes over time because the density of matter declines as the Universe ages and continues to expand. Over long periods of time, the number of stars declines. Stars like the Sun last for about 10 billion years before they run out of hydrogen in their cores and cannot make helium to generate energy and keep the star stable. More massive stars "burn out" quicker, and less massive ones last longer. The lowest mass stars have only 10 percent of the mass of the Sun. These stars are much less luminous than the Sun and can generate nuclear energy in their cores for about 10 trillion years, over a hundred times longer than the current age of the Universe. Only stars more than half the mass of the Sun have hot enough cores to make carbon, and they last less than 20 billion years.

As cosmic time marches on, the rate of star formation declines, and eventually stars massive enough to make carbon cease to form. This will truly be a watershed moment in the carbon budget of the Universe. Carbon has been made for billions and billions of years, and its total cosmic abundance has grown. In the 13.8 billion years since the Big Bang, the amount of carbon atoms in the Universe increased from zero to about 0.03 percent of the number of hydrogen atoms. This amazing element is still rare compared to hydrogen, and it will never be much more abundant than it is now, but its accumulation has made us possible and yielded all of the wonders described in this book. When the stars that make it no longer form, the universal abundance of carbon will decline as the Universe progresses to the difficult-to-imagine future that will occur if it just continues to expand forever. (The expansion rate is even accelerating!) Carbon, as well as other forms of matter, will wane. Some of the loss is just because stars, planets, interstellar matter, and even cosmic rays can fall into black holes, where the force of gravity is so strong that light cannot escape. All matter that falls into a black hole is consumed by a genuine monster of nature. Atoms and even nuclear particles lose their identity as they contribute to the mass of the growing monster. Galaxies like our own contain massive central black holes that, like cosmic vacuum cleaners, assimilate and destroy materials that get too close.

The Universe in distant time will run down. As time marches toward eternity, the Universe will become a strange, dark place without glowing stars and filled with contents that become increasingly isolated from each other. It is very hard for us humans to relate to the concept of forever. In the very distant future, the lowest-mass stars alone will remain for a while, and only as dim proxies of their former glory. As the decline proceeds, the dimly glowing stellar sources of light will slowly fade and eventually reach an end state of burned-out stars or worse. As the more massive stars evolve, they explode, shedding mass, leaving cores that become black holes, neutron stars, or white dwarfs, depending on the star's initial mass. Stars that are more than about twenty times the mass of the Sun evolve to become core-collapse supernovae and produce black holes. Lesser-mass stars become neutron stars or white dwarfs. Matter that is blown off into space by explosions goes into the interstellar medium. The escaped matter will contain and preserve carbon as either atoms, molecules, or small grains of cosmic dust. Some of this material can exist in vacuous space essentially forever.

Carbon that stays in the cores of massive stars is destroyed by the formation of either a black hole or a neutron star. In ultra-dense neutron stars, most previously existing atoms are totally transformed, leaving a nearly pure mass of neutrons, a huge, extraordinary mass of nuclear matter. Neutron stars have stellar masses but are incredibly small, dense bodies. They are the densest bodies in the Universe and are only about 10 kilometers in diameter. Our galaxy contains about a billion of them. They form when interior conditions in massive stars become so extreme that protons and electrons combine to form neutrons, particles with no electrical charge. Stars more than 1.4 times the mass of the Sun (known as the Chandrasekhar limit) explode, but less-massive ones can form white dwarf stars that can last indefinitely. With masses below the Chandrasekar limit, the density becomes so high that the electrons fill all available energy states and enter a "degenerate state," where quantum mechanical effects allow them to provide enough pressure to support the star against its own gravity. This state is independent of temperature, and the star can be stable for essentially forever, no matter how cold it gets.

White dwarfs also form as the end state of stars that do not collapse to form black holes or neutron stars and have less mass than the Chandrasekhar limit. They are derived from stars that went through their full nuclear element production phase. Over 97 percent of the present inventory of stars will end up as white dwarfs, and these stellar mass bodies, compressed to the size of Earth, will remain forever as tombstones of the stars that preceded them. The most common white dwarfs are dominated by carbon and oxygen. When they form, their surfaces can exceed one hundred thousand degrees centigrade, but with their nuclear reactions exhausted, they slowly cool down and become fainter. Eventually they will become too faint to detect and can be considered black dwarfs. The Universe is young enough that invisible black dwarfs have not yet formed. White dwarfs cool to the point where their interiors crystallize and can form diamonds. Just imagine our wonderful element carbon finally ending up as giant diamonds in the cold interiors of invisible black dwarf stars.

To think about cosmic evolution and the fate of carbon in earthly terms, let us consider what will happen to the carbon atoms that the Lavoisiers burned up in their famous diamond experiment in 1772. They burned diamond with sunlight focused to a point by an impressively large "burning lens" and found that the invisible gas that was produced exactly matched the mass of diamond that burned up and vanished. These landmark experiments, which led to the naming of carbon and oxygen, greatly advanced our understanding of atoms, and marked the emergence of chemistry as a scientific enterprise.

Some of the carbon dioxide from Lavoisier's sealed combustion experiments ended up back in the atmosphere. Gas molecules are tiny, and there were almost a trillion times a trillion of them from the experiments. Where did they go? Most dissolved in the ocean, some were taken out of the air to form plants, some went into shells, and some went into forming carbonate rocks.

Many remained in the atmosphere, and you breathe carbon from Lavoisier's experiments in and out all the time, even after two centuries. Your bodies actually contain some carbon atoms from Lavoisier's diamond. Atoms are really small! Twelve grams of diamond, a little less than a teaspoon, contain 6.02×10^{23} atoms (known as Avogadro's number). If uniformly spread across Earth's surface, each square meter would contain a billion carbon atoms from Lavoisier's diamond.

What about the future? The Earth's ongoing carbon cycles will continue until the Sun finally gets so bright that both its biologically driven and its plate tectonic–driven carbon cycles will cease. Eventually, our oceans and atmosphere will be lost, and the carbon in them will stream into space. In space, it will become ionized and join the outward flow of solar wind that drives released carbon atoms out of the solar system at speeds of hundreds of kilometers per second. No matter how nasty the conditions become, a major fraction of our planet's carbon will remain in its silicate mantle and the iron core. If Earth survives the late-stage evolution of the Sun, this carbon will remain forever trapped in our ever-cooling frozen planet orbiting a dead star.

As bad as this sounds, the Earth's actual fate is probably worse, or at least quite different, depending on your opinion of what might be better or worse in the quite distant future. Earth formed and has resided in the solar system's prime real-estate district, where solar heating is just right so that our oceans have never totally frozen or vaporized away. This is, of course, the aptly named habitable zone. As we have seen, there are dire consequences of living so close to a star that it provides adequate warmth and light. When the Sun enters the red giant stage, in the last 10 percent of its lifetime, it will brighten by several thousand times and expand to a size that almost reaches Earth's orbit. All planets that flourish in an HZ ultimately get fried when their stars enter old age. Mars and the planets beyond can survive the heat of the red giant Sun, but Earth probably will not. Earth will probably end up going into the Sun, and Venus and Mercury certainly will. The Earth is just at the edge of the danger zone, but it is believed that Earth's gravity will raise a tidal "bump" on the greatly swollen Sun that will pull on the Earth and cause it to spiral inward until it enters the Sun's atmosphere. Earth presently has an orbital speed of 30 km/19 miles per second, and at this great speed friction with gas at the edge of the Sun will totally vaporize our planet to individual atoms. The temperature of the surface of the red giant Sun is actually cooler than the Earth's present interior—Earth will not be killed by intense heat and light from the Sun but by energy from high-speed collisions with atoms in the Sun's atmosphere.

Is this the end of the story? No! When Earth makes its fiery entry into the Sun, the temperatures will rise to the point where all molecules will be broken into individual atoms. This event will happen when the red giant Sun is so bright and so enlarged that a significant fraction of its mass is streaming off into space. This ejection of mass by dying stars is commonly observed astronomically, and the ejected material has produced some of the most spectacular images ever taken with telescopes. Our planet's precious carbon atoms will most likely all be ejected back into space, along with the outward flow of solar material escaping our star. It will join the interstellar medium in our Milky Way. Some of the atoms will end up in black holes, and some will be involved in forming new stars and planets. Some can even be swept out of our Galaxy and either enter another galaxy or become stranded in the vacuous intergalactic medium.

It is intriguing to imagine that at least some of our treasured hoard of wonderful carbon atoms can be cycled for trillions of years into multiple future stars, planets, and perhaps even organisms before ultimately ending up in the deep time of future evolution where cosmic evolution heads toward infinity, a direction that most humans have a difficult time imagining. In our view as astronomers, it is comforting to expect that Earth's carbon atoms will continue to be recycled for quite an extended time. (If we had been born on Mars, our precious carbon atoms would have been inside a body that would survive the red giant stage but would then have been locked up nearly forever inside a body cooling slowly toward absolute zero temperature.)

Astronomers deal with billions of years on nearly a daily basis and become hardened to the concept of deep time. But it is entirely natural that many people are unsettled when thinking about the vastness of the Universe and the endless progression of time. We humans have no control over the longest-term paths of nature, but we can take solace and considerable satisfaction that we have reached a point where we are able to understand the workings of nature well enough to comprehend the basic details of fascinating subjects such as the element carbon, both in our cosmic past and in the distant future. We owe this comforting attitude to watershed contributions from the likes of the Lavoisiers, Fred Hoyle, Louis Pasteur, Edwin Hubble, Rosalind Franklin, Clair Patterson, Joseph Fourier, Harry Kroto, and many others. Perhaps the best way to contemplate the distant fate of carbon and everything else in the Universe is just to ponder this famous quote, most commonly attributed to Woody Allen or Steven Hawking, "Eternity is an awful long time, especially towards the end."

NOTES

Chapter 1: The Discovery, Origin, and Dispersal of Carbon

1. Uranium is often considered to be the heaviest naturally occurring element, but it is not. Heavier transuranium elements are made in the Universe, and we know that natural plutonium (ninety-four protons) was in the solar system at the time that planets were forming, although it has since decayed. Very minor amounts of plutonium and perhaps the transuranium elements neptunium and californium are also believed to be naturally produced in present-day uranium ores.

2. The appellation "Big Bang" was first uttered in the 1950s by a nonbeliever, as a joke that stuck. The nonbeliever was Sir Fred Hoyle, a famous and essential British scientist who will appear often in these pages. Hoyle thought the Universe was forever, with no beginning and no end. The idea that the Universe began as a singular event was suggested in 1931 by Georges Lemaître, a cosmologist and Catholic priest.

3. Gamow had a fondness for puns, which he indulged in the author list for the 1948 epochal paper, "The Origin of the Chemical Elements," published in the journal *Physical Review*. The authors were Alpher, Gamow, and a third author, who was not involved in the research—Hans Bethe. His name was added by Gamow to make a pun, so the paper could have as its authors Alpher, Bethe, and Gamow, a play on the first three letters of the Greek alphabet, α, β, and γ (alpha, beta, gamma). The research was in fact Alpher's PhD thesis.

4. Many of the pretty astronomical images you see in pictures and posters or online show star-forming regions: the brilliant colors in these clouds and nebulae are from their many atoms, ions, and molecules emitting light at different visible wavelengths. Notably, hydrogen, the most abundant element, has a strong emission line in the red portion of the color spectrum.

5. Despite her growing reputation and PhD, her title at Harvard was "technical assistant." Women still had a secondary place in academia and were not allowed to be Harvard professors. Only later, in 1934, was she appointed as a professor—and chair of the astronomy department at Harvard! This woman, who revolutionized our view of the cosmos, was the thesis adviser for Paul Hodge at Harvard, who was thesis

adviser for many astronomers, including one of the authors of this book (DB). She also gave guiding thesis advice to Harvard student Jesse Greenstein, who was the thesis adviser of a Caltech astronomy student named George Wallerstein, who was the thesis adviser of the first author of this book (TPS).

Chapter 2: The Chemistry of Carbon: Why Is It So Special?

1. In the United States, old-school chemistry sets became extinct due to new regulations, such as the Federal Hazardous Substances Labeling Act (1960), the Toy Safety Act (1969), the Consumer Product Safety Commission (1972), and the Toxic Substances Control Act (1976).

2. Instead of "orbits," physicists usually say "orbitals" or "shells" to indicate that the orbits are not in one plane; instead, they may be spherical, shaped like lobes, or appear in other configurations. Furthermore, in the quantum physics domain, we don't say exactly where an electron is at any given time; we assign probabilities instead. The result is that we refer to the "electron cloud." The closer you look, the messier the physics.

3. There is also metallic bonding. As the name implies, this occurs in metals where outermost electrons are free to move. This is how electricity works in conducting materials such as copper wires.

4. The Ig Nobel Prize for quirky discoveries is awarded yearly by Harvard and MIT and is a satire of the real Nobel Prizes. Geim won both.

5. Buckminster Fuller was a futurist and at the same time a true Renaissance man: architect, inventor, and engineer. He invented the geodesic dome, a very stable arrangement of flat panels. The new sixty-carbon atom molecule was a natural name in honor of Fuller.

Chapter 3: Carbon on Earth and in the Solar System

1. Kepler was also an astrologer—the difference with astronomy was hardly noticed at the time. Maybe astrologists made more money than philosophers, as scientists were called then.

2. A half-life is what it sounds like. For example, if you have 1 kg of the ^{14}C at first, in 5,730 years you will have one-half of 1 kilogram. Note that the leftover ^{14}C is still radioactive. In another 5,730 years, there will be one-quarter of 1 kg of radioactive ^{14}C, in a further 5,730 years one-eighth, and so on.

3. Some people don't like Pluto being downgraded from a planet to a dwarf planet, especially in Las Cruces, New Mexico, where Tombaugh was a faculty member at New Mexico State University, and where Pluto Day is celebrated every February 18, the anniversary of the announcement of Pluto's discovery.

Chapter 4: Carbon and Life on Earth and Elsewhere

1. Haldane was an interesting character. From a background of English nobility and wealth, not to mention the best schools, he gradually became a socialist and then a communist, never completely disavowing Hitler or Stalin.

2. By far the majority of interstellar molecules are found by their radio emission spectra. Of the 250 (and counting) species discovered so far, most are carbon based. A prominent one is ethyl alcohol, better known as booze.

3. Much earlier (1726) in Jonathan Swift's *Gulliver's Travels*, Mars was depicted with two tiny moons orbiting it, which inspired many science fiction novels—but that was just luck.

4. The timing was not accidental. The spacecraft first went into Martian orbit, and the landers touched down soon after, in 1976—almost exactly on the two hundredth birthday of the US.

Chapter 5: Carbon in the Milky Way

1. The Hubble Space Telescope (HST) was envisioned long before the first artificial satellites were in orbit. In 1946, Lyman Spitzer proposed a telescope above the atmosphere that could make the capturing of images and spectra in ultraviolet wavelengths possible. The HST is named for Edwin Hubble, the astronomer who discovered the expanding universe. It could have been the SST, named for Spitzer, but Spitzer was still alive in 1990, when the space telescope was launched, and NASA had a policy that spacecraft could not be named after living people. After Spitzer's death, in 1997, another orbiting infrared telescope was named for him, the Spitzer Space Telescope.

2. Positive ions are electrically charged atoms with one or more electrons missing. Neutral carbon atoms have six electrons, but in astronomical environments, ultraviolet light or high temperature can cause electrons to be lost, unbound. Neglecting negative ions, carbon has a total of six ion stages, plus neutral atoms, making seven possible kinds of carbon. Astronomers label the stages C I (neutral atoms) to C VII (atoms with all six electrons missing, only present in extremely hot gas). Chemists use a different terminology: C, C^+, C^{++}, C^{+++}, C^{++++}, but the highly ionized carbon atoms are not seen in normal laboratory environments.

3. All detected interstellar molecules are listed here: http://www.astrochymist .org/astrochymist_ism.html and https://cdms.astro.uni-koeln.de/cdms/portal/. Note that most of them contain carbon.

4. As noted previously, PAH stands for polycyclic aromatic hydrocarbon. We'll see them again later in this chapter.

5. This is the same Barnard whose observations of dark interstellar clouds were described above. Not everyone has a star named for them; you have to do something special, like discovering the interstellar medium!

Chapter 6: What Is Carbon Good For?

1. Oil spills have become legends, even to the point of inspiring folk song writers. Several spills provoked the folksinger Steve Forbert to record "The Oil Song" in 1979: "And it's oil, oil, drifting to the sea. / And it's oil, oil. / Don't buy it at the station. / You can get it for free. / Just come to the shoreline where the water used to be."

2. Because of their high metabolic rate, rapid breathing, and sensitivity to carbon monoxide, canaries used to be carried into mines to warn the miners of poison gas buildup by dying before miners succumbed to it. Fortunately, canaries have now been replaced by electronic monitors.

3. Could you lose weight by eating celery? Celery has a high level of cellulose, so high that the energy gain from eating celery is less than the energy it takes to digest it. Most other foods, of course, exhibit the opposite of this trend.

4. Teflon even has a political definition. In the 1980s, Ronald Reagan was sometimes called the "Teflon president" because he narrowly escaped scandals a few times; they couldn't stick.

Chapter 7: Diamonds

1. A list of the largest rough diamonds can be found at https://en.wikipedia.org/wiki/List_of_largest_rough_diamonds.

2. The first major diamond field is located near the town of Kimberley in South Africa, hence "kimberlite" as the name of the diamond-bearing rock.

3. That's equivalent to more than $15,000,000 in 2023 (https://www.in2013dollars.com/us/inflation/1872?amount=650000).

4. The still widely used Mohs hardness scale was designed in 1812 by German mineralogist Friedrich Mohs. If one mineral makes a scratch on another, it means it is harder than the other. This scale is ordered but very nonlinear. On an absolute scale, diamonds (10) are nearly four times as hard as sapphire (9) and about fifteen hundred times as hard as talc (1).

Chapter 8: The Atmosphere, Climate, and Habitability

1. As an aside, it is interesting to note that Mars also has a carbon short-term cycle, but not for the same reasons that Earth does. Over the Martian year, it gets so cold in polar regions that 40 percent of the atmospheric CO_2 freezes out as dry ice. It then sublimates back into the air as Martian winter turns to spring.

Chapter 9: Carbon Out There

1. World War II was going on when Baade was observing the sky from Mount Wilson, high above Pasadena, California, and Southern California had its lights turned off in fear of bombing from Japanese ships. Because of his divided heritage, Swiss and American, Baade was barred from serving in the American army—and he had many dark sky nights to see Andromeda from the 100-inch Hooker telescope. His Mount Wilson work also led to a considerable understanding of the distances to other galaxies.

2. "3C" indicates it is from the third Cambridge catalog.

3. The constellation Sagittarius contains a black hole. It is in the center of our Galaxy, and the exact location of the black hole is called Sgr A*. We see only the gas and stars surrounding it, and only by observing it in radio and infrared wavelengths. All visible light from Sgr A* is blocked by intervening gas and dust. By analyzing the motions of stars around the black hole, astronomers can deduce its mass. At 3.6 million times the mass of the Sun, it is a super-massive black hole, but, at least at present, it is not a quasar because its mass or mass inflow is not great enough to generate a quasar level of power. In May 2022, the Event Horizon Telescope (EHT) consortium was able to use radio telescopes across the Earth to image the black hole Sgr A*. The fantastic doughnut-like image was made by radio waves whose paths were bent by the gravity of the massive black hole.

REFERENCES

Chapter 1

1.1 West, John B. "The Collaboration of Antoine and Marie-Anne Lavoisier and the First Measurements of Human Oxygen Consumption." *American Journal of Physiology—Lung Cellular and Molecular Physiology* 305, no. 11 (2013): L775–85.

1.2. Eagle, Cassandra T., and Jennifer Sloan. "Marie Anne Paulze Lavoisier: The Mother of Modern Chemistry." *Chemical Educator* 3 (1998): 1–18.

1.3 Bohning, James J. "The Chemical Revolution." *American Chemistry Life* (1999). https://www.acs.org/content/dam/acsorg/education/whatischemistry /landmarks/lavoisier/antoine-laurent-lavoisier-commemorative-booklet.pdf.

1.4 Wertime, T. A. "The Discovery of the Element Carbon." *Osiris* 11 (1954): 211–20.

1.5 Harwit, Martin. "Ralph Asher Alpher." *Physics Today* 60, no. 12 (2007): 67.

1.6 Hoyle, Fred. *Home Is Where the Wind Blows: Chapters from a Cosmologist's Life.* Mill Valley, CA: University Science Books, 1994.

1.7 Wolchover, Natalie. "A Primordial Nucleus behind the Elements of Life." *Quanta Magazine*, December 4, 2012.

1.8 Eid, Mounib El. "The Process of Carbon Creation." *Nature* 433, no. 7022 (2005): 117–19.

1.9 Burbidge, E. Margaret, Geoffrey Ronald Burbidge, William A. Fowler, and Fred Hoyle. "Synthesis of the Elements in Stars." *Reviews of Modern Physics* 29, no. 4 (1957): 547–650.

1.10 Burbidge, Geoffrey. "Hoyle's Role in B^{2}FH." *Science* 319, no. 5869 (2008): 1484.

1.11 Kippenhahn, Rudolf, Alfred Weigert, and Achim Weiss. *Stellar Structure and Evolution.* Berlin, Heidelberg: Springer Berlin Heidelberg, 2012.

1.12 Gelling, Natasha. "The Women Who Mapped the Universe and Still Couldn't Get Any Respect." *Smithsonian Magazine.* September 18, 2013. https://www .smithsonianmag.com/history/the-women-who-mapped-the-universe-and -still-couldnt-get-any-respect-9287444/.

1.13 Sobel, Dava. *The Glass Universe.* New York: Penguin Books, 2017.

Chapter 2

2.1   Shuttleworth, Martyn. "Islamic Alchemy." Explorable. November 23, 2010. https://explorable.com/islamic-alchemy.

2.2   Elsergany, Ragheb. "Muslims and the Invention of Chemistry." Islamastory.com. https://islamstory.com/en/artical/23561/Muslims-Invention-Chemistry.

2.3   Scerri, Eric R. *The Periodic Table: A Very Short Introduction.* Vol. 289. Oxford, UK: Oxford University Press, 2011.

2.4   Scerri, Eric R. *The Periodic Table.* New York, Oxford: Oxford University Press, 2006.

2.5   Massimi, Michela. "Pauli's Exclusion Principle: The Origin and Validation of a Scientific Principle." In *Pauli's Exclusion Principle: The Origin and Validation of a Scientific Principle.* Cambridge: Cambridge University Press, 2005.

2.6   Pauli, W. "Über den Zusammenhang des Abschlusses der Elektronengruppen im Atom mit der Komplexstruktur der Spektren." *Zeitschrift für Physik* 31 no. 1 (1925): 765–83.

2.7   Dingle, Adrian. "Explainer: What Are Chemical Bonds?" ScienceNewsExplores. April 29, 2021. https://www.snexplores.org/article/explainer-what-are-chemical -bonds.

2.8   Coapinto, John. "Material Question. Graphene May Be the Most Remarkable Substance Ever Discovered. But What's It For?" *The New Yorker* 22 (2014): 29.

2.9   Kroto, Harold W., James R. Heath, Sean C. O'Brien, Robert F. Curl, and Richard E. Smalley. "C_{60}: Buckminerfullerene" *Nature* 318, no. 6042 (1985): 162–63.

2.10 Ibid.

2.11 Krätschmer, Wolfgang, and Donald R. Huffman. "Production and Discovery of Fullerites: New Forms of Crystalline Carbon." *Philosophical Transactions of the Royal Society of London. Series A: Physical and Engineering Sciences* 343, no. 1667 (1993): 33–38.

Chapter 3

3.1   Hazen, Robert. "How Old Is Earth, and How Do We Know?" *Evolution:Education and Outreach* 3 (2010), 198–205.

3.2   Reilly, Lucas. "The Most Important Scientist You've Never Heard Of." Mental Floss. May 17, 2017. https://www.mentalfloss.com/article/94569/clair -patterson-scientist-who-determined-age-earth-and-then-saved-it.

3.3   Croswell, Ken. "An Elemental Problem with the Sun." *Scientific American.* July 1, 2020. https://www.scientificamerican.com/article/an-elemental-problem-with -the-sun/.

3.4 Gail, Hans-Peter, Mario Trieloff. "Spatial Distribution of Carbon Dust in the Early Solar Nebula and the Carbon Content of Planetisimals." *Astronomy and Astrophysics* 606, no. A16 (2017). https://www.aanda.org/articles/aa/full_html /2017/10/aa30480-17/aa30480-17.html.

3.5 Hirose, Kei, Bernard Wood, and Lidunka Vočadlo. "Light Elements in the Earth's Core." *Nature Reviews Earth and Environment* 2, no. 9 (2021): 645–58.

3.6 Fischer, Rebecca A., Elizabeth Cottrell, Erik Hauri, Kanani K. M. Lee, and Marion Le Voyer. "The Carbon Content of Earth and Its Core." *Proceedings of the National Academy of Sciences* 117, no. 16 (2020): 8743–49.

3.7 Ahrer, Eva-Maria, Lili Alderson, Natalie M. Batalha, Natasha E. Batalha, Jacob L. Bean, et al. "Identification of Carbon Dioxide in an Exoplanet Atmosphere." arXiv preprint arXiv:2208.11692 (2022).

Chapter 4

4.1 Bottke, William F., and Marc D. Norman. "The Late Heavy Bombardment." *Annual Review of Earth and Planetary Sciences* 45 (2017): 619–47.

4.2 Gözen, Irep, Elif Senem Köksal, Inga Põldsalu, Lin Xue, Karolina Spustova, Esteban Pedrueza-Villalmanzo, Ruslan Ryskulov, Fanda Meng, and Aldo Jesorka. "Protocells: Milestones and Recent Advances." *Small* 18, no. 18 (2022): e2106624. https://www.doi.org/10.1002/smll.202106624.

4.3 Weiss, Madeline C., Martina Preiner, Joana C. Xavier, Verena Zimorski, and William F. Martin. "The Last Universal Common Ancestor between Ancient Earth Chemistry and the Onset of Genetics." *PLoS Genetics* 14, no. 8 (2018): e1007518.

4.4 Trefil, James, Harold J. Morowitz, and Eric Smith. "The Origin of Life: A Case Is Made for the Descent of Electrons." *American Scientist* 97, no. 3 (2009): 206–13.

4.5 Olejarz, Jason, Yoh Iwasa, Andrew H. Knoll, and Martin A. Nowak. "The Great Oxygenation Event as a Consequence of Ecological Dynamics Modulated by Planetary Change." *Nature Communications* 12, no. 1 (2021): 1–9.

4.6 Kopp, Robert E., Joseph L. Kirschvink, Isaac A. Hilburn, and Cody Z. Nash. "The Paleoproterozoic Snowball Earth: A Climate Disaster Triggered by the Evolution of Oxygenic Photosynthesis." *Proceedings of the National Academy of Sciences* 102, no. 32 (2005): 11131–36.

4.7 Ward, P. D., and D. Brownlee. "*Rare Earth: Why Complex Life Is Uncommon in the Universe.*" New York: Copernicus Books, 2000.

4.8 Schulte, Peter, Laia Alegret, Ignacio Arenillas, José A. Arz, Penny J. Barton, Paul R. Bown, Timothy J. Bralower, et al. "The Chicxulub Asteroid Impact and Mass Extinction at the Cretaceous-Paleogene Boundary." *Science* 327, no. 5970 (2010): 1214–18.

4.9 Kasting, James. *How to Find a Habitable Planet*. Princeton, NJ: Princeton University Press, 2010.

4.10 Parker, Eric T., James H. Cleaves, Aaron S. Burton, Daniel P. Glavin, Jason P. Dworkin, Manshui Zhou, Jeffrey L. Bada, and Facundo M. Fernández. "Conducting Miller-Urey Experiments." *JoVE (Journal of Visualized Experiments)* 83 (2014): e51039.

Chapter 5

5.1 McCray, Richard, and Theodore P. Snow. "The Violent Interstellar Medium." *Annual Review of Astronomy and Astrophysics* 17 (1979): 213–40.

5.2 Kudritzki, Rolf-Peter, and Joachim Puls. "Winds from Hot Stars." *Annual Review of Astronomy and Astrophysics* 38, no. 1 (2000): 613–66.

5.3 Cox, Nick L. J. "The Diffuse Interstellar Bands: An Elderly Astro-Puzzle Rejuvenated." *The Molecular Universe, Proceedings of the International Astronomical Union, IAU Symposium* 280 (2011): 162–76.

5.4 Ehrenfreund, Pascale, and Jan Cami. "Cosmic Carbon Chemistry: From the Interstellar Medium to the Early Earth." *Cold Spring Harbor Perspectives in Biology* 2, no. 12 (2010): a002097.

5.5 Borucki, William J. "KEPLER Mission: Development and Overview." *Reports on Progress in Physics* 79, no. 3 (2016): 036901. https://www.doi.org/10.1088/0034-4885/79/3/036901.

5.6 Young, E. D., A. Shahar, and H. E. Schlichting. "Earth Shaped by Primordial H_2 Atmospheres." *Nature* 616 (2023): 306–11.

5.7 Allen-Sutter, Harrison, E. Garhart, K. Leinenweber, V. Prakapenka, E. Greenberg, and S-H. Shim. "Oxidation of the Interiors of Carbide Exoplanets." *Planetary Science Journal* 1, no. 2 (2020): 39.

5.8 Burchell, Mark J. "W(h)ither the Drake Equation?" *International Journal of Astrobiology* 5, no. 3 (2006): 243–50.

5.9 Greshko, Michael. "Frank Drake, Pioneer in the Search for Alien Life, Dies at 92." *National Geographic.* September 2, 2022. https://www.nationalgeographic.com/science/article/frank-drake-pioneer-in-the-search-for-alien-life-dies-at-92.

Chapter 6

6.1 Brown, Malcolm W. "Alcohol-Laden Cloud Holds the Story of a Star." *New York Times*, May 30, 1995. https://www.nytimes.com/1995/05/30/science/alcohol-laden-cloud-holds-the-story-of-a-star.html.

6.2 Weart, Spencer. "Roger Revelle's Discovery." American Institute of Physics. August 2022. https://history.aip.org/climate/Revelle.htm.

6.3 Bennett, Matthew R., David Bustos, Jeffrey S. Pigati, Kathleen B. Springer, Thomas M. Urban, Vance T. Holliday, Sally C. Reynolds, et al. "Evidence of Humans in North America during the Last Glacial Maximum." *Science* 373, no. 6562 (2021): 1528–31.

6.4 Miyake, Fusa, Kentaro Nagaya, Kimiaki Masuda, and Toshio Nakamura. "A Signature of Cosmic-Ray Increase in AD 774–775 from Tree Rings in Japan," *Nature* 486, (2012): 240–42.

Chapter 7

7.1 Smith, Evan M., Steven B. Shirey, and Wuyi Wang. "The Very Deep Origin of the World's Biggest Diamonds." *Gems and Gemology* 53, no. 4 (2017): 388–403.

7.2 Smith, Evan M., Steven B. Shirey, Stephen H. Richardson, Fabrizio Nestola, Emma S. Bullock, Jianhua Wang, and Wuyi Wang. "Blue Boron-Bearing Diamonds from Earth's Lower Mantle." *Nature* 560 (2018): 84–87.

7.3 Smith, Evan M., and Fabrizio Nestola. "Super-Deep Diamonds: Emerging Deep Mantle Insights from the Past Decade." *Mantle Convection and Surface Expressions* (2021): 179–92.

7.4 Pappas, Stephanie. "Ultra Rare Diamond Suggests Earth's Mantle Has an Ocean's Worth of Water." *Scientific American*. September 26, 2022. https://www.scientificamerican.com/article/oceans-worth-of-water-hidden-deep-in-earth-ultra-rare-diamond-suggests/.

7.5 Boissoneault, Lorraine. "The True Story of the Koh-i-Noor Diamond—and Why the British Won't Give It Back." *Smithsonian Magazine*. August 30, 2017. https://www.smithsonianmag.com/history/true-story-koh-i-noor-diamondand-why-british-wont-give-it-back-180964660/.

Chapter 8

8.1 Weart, Spencer. "The Discovery of Global Warming." American Institute of Physics. April 2022. https://history.aip.org/climate/index.htm.

8.2 Kasting, James F., and Janet L. Siefert. "Life and the Evolution of Earth's Atmosphere." *Science* 296, no. 5570 (2002): 1066–68.

8.3 Crutzen, Paul J. "Geology of Mankind: The Anthropocene." *Nature* 415, no. 23 (2002).

8.4 Howe, Joshua P. "This Is Nature; This Is Un-nature: Reading the Keeling Curve." *Environmental History* 20, no. 2 (2015): 286–93.

8.5 Summerhayes, Colin P. *Paleoclimatology: From Snowball Earth to the Anthropocene*. Hoboken, NJ: John Wiley & Sons, 2020.

8.6 Steffen, Will, Johan Rockström, Katherine Richardson, Timothy M. Lenton, Carl Folke, Diana Liverman, Colin P. Summerhayes, et al. "Trajectories of the Earth System in the Anthropocene." *Proceedings of the National Academy of Sciences* 115, no. 33 (2018): 8252–59.

8.7 Jackson, Roland. "Who Discovered the Greenhouse Effect?" The Royal Institution. 2019. https://www.rigb.org/explore-science/explore/blog/who-discovered -greenhouse-effect.

8.8 Weart, Spencer. "The Discovery of Global Warming." American Institute of Physics. April 2022. https://history.aip.org/climate/index.htm.

8.9 Stefaniuk, Damian, Marcin Hajduczek, James C. Weaver, Franz J. Ulm, and Admir Masic. "Cementing CO_2 into C-S-H: A Step toward Concrete Carbon Neutrality." *PNAS Nexas* 2, no. 3 (2023): 1–5. https://academic.oup.com /pnasnexus/article/2/3/pgad052/7089570.

8.10 Wei-Haas, Maya. "We Are Missing at Least 145 Carbon-Bearing Minerals, and You Can Help Find Them." *Smithsonian Magazine*. December 17, 2015. https:// www.smithsonianmag.com/science-nature/we-are-missing-145-carbon -bearing-mineral-you-can-help-find-them-180957575/.

8.11 Ward, Peter Douglas. *Rivers in Time: The Search for Clues to Earth's Mass Extinctions*. New York: Columbia University Press, 2000.

Chapter 9

9.1 Beers, Timothy C., and Norbert Christlieb. "The Discovery and Analysis of Very Metal-Poor Stars in the Galaxy." *Annual Review of Astronomy and Astrophysics* 43 (2005): 531–80.

9.2 Drake, Nadia. "The Most Ancient Galaxies in the Universe Are Coming into View." *National Geographic*. January 26, 2023. https://www.nationalgeographic .com/magazine/article/nasa-jwst-most-ancient-galaxies-in-universe-coming -into-view.

9.3 Overbye, Dennis. "Who Will Have the Last Word on the Universe?" *New York Times*, May 2, 2023.

9.4 O'Callaghan, Jonathan. "At Last, Astronomers May Have Seen the Universe's First Stars," *Scientific American*. June 13, 2023. https://www.scientificamerican .com/article/at-last-astronomers-may-have-seen-the-universes-first-stars/.

9.5 Spilker, Justin S., Kedar A. Phadke, Manual Aravena, et al. "Spatial Variations in Aromatic Hydrocarbon Emission in a Dust-Rich Galaxy." *Nature*. June 5, 2023.

INDEX